AF870580

Cold War: A Military Memoir

1955-1978

Captain Frank A. Bausch

United States Navy (Retired)

Also by Frank A. Bausch

Reasoning Together: Essays on current economic, political and social issues prefaced with a tract on what constitutes 'right thinking.'

Bridges Crossed (and some not): Pictorial travelogue with history and construction details of selected North American rail and road bridges.

The Joy of Treason: A novel about a couple who ostensibly betray their country but do so in a way that minimizes the risk and maximizes the fun.

Bausch Family Chronicle: Illustrated history of the Bausch family with emphasis on the latest generation and their immediate forebears.

Travels with Mary: Recounting of rail and automobile trips illustrated with numerous photos covering 55 years of extensive travel in the US and Canada.

Nothing like Losing: Retrospective commentary on 12 years of participating in small town municipal government (available April 2013).

About the Author: Frank A. Bausch lives with his wife in Caswell Beach NC; they have six grown children living in various parts of the US. He is a retired Navy Captain with squadron, aircraft carrier, fleet staff and Pentagon experience. After leaving the Navy, he worked for CIA and DIA and then spent ten years as a Senior Member of the Technical Staff for E-Systems (Raytheon). He holds degrees from St. John's, NY and the Univ. of Maryland.

Cover Photo: Two F/A-18C Hornets silhouetted by lightning on the flight deck of the aircraft carrier USS John C. Stennis in the Pacific Ocean (Mass Communication 2nd Class Elliott Fabrizio/US Navy).

Type Face: Comic Sans is a casual script modeled on fonts used in American comic books for several decades and designed by Vincent Connare. *Sans* is short for sans-serif.

ISBN-10: 098535173X
ISBN-13: 978-0-9853517-3-1 (softcover)

Table of Contents

Introduction	3
Chapter 1—Naval Training—OCS/AGO/PIC (1955)	7
Chapter 2—Pacific Fleet (1956-1959)	11
Chapter 3—Navy PIC (1960-1963)	29
Chapter 4—Atlantic Fleet (1963-1967)	39
Chapter 5—Defense Intelligence School (1967-1968)	51
Chapter 6—Defense Intelligence Agency (1968-1971)	55
Chapter 7—AFAITC Lowry AFB (1971-1973)	69
Chapter 8—Defense Intelligence Agency (1973-1976)	81
Chapter 9—Intelligence Community Staff (1976-1978)	93
Appendix A—Awards & Citations	105
Appendix B—Miscellaneous Photos/Documents	111

Introduction

In July 2008, at our annual 'beach week' family get-together, my oldest daughter and I were talking about higher education and in passing I mentioned my efforts in the navy during the early 1970s towards getting a PhD. This turned out to be news to her, which made me think about the many other things my wife and children don't know about my military experience. That would be something I would correct if I had to do it all over again, but even had I been more liberal with sharing happenings at work, my family probably could not have grasped the scope and complexity of what I was involved in. The 24 years of my life chronicled in this memoir is an attempt to remedy that shortcoming. Written primarily for family and friends, I trust it will give them and other readers some insight into what I did at the office.

The Environment

What follows covers a period that ended more than thirty years ago. Looking back, it seems like a time not too dissimilar from the present. A clearly perceived threat to the country existed; now it is Islamic terrorism represented by Al Qaeda, then it was international communism fostered by Soviet imperialism. Early in the period, the military was actively engaged in warfare; substitute Korea and Vietnam for Afghanistan and Iraq to see a rough parallel. A one time co-worker, Bob Sorley, author of a *Better War,* described the Vietnam era this way: "Those years constituted one of the most complex and difficult periods the country, and its armed forces, has ever gone through—a limited war within the larger Cold War within a global cultural revolution and ultimately (in respect to Vietnam) a failed endeavor."

As for differences, one was the scale of active conflict: almost 60,000 US servicemen died in Vietnam compared to something over 4000 in Iraq. But there was another big difference; the apprehension caused by the possibility of nuclear war dwarfed that of terrorist attacks today. Terrorism is what generally happens somewhere else; nuclear war in those days was a potential end to life on earth for everyone. National security policy embracing the concept of nuclear deterrence was very nuanced but what was very straightforward was the commitment by the military in fighting this cold war.

The depth of this involvement was most evident in the military elite force structure, e.g. strategic aviation, submariners, special forces, etc. But the cold war conflict was also every bit as intense for those who had to assess just how dangerous the threat was. As you will read, fairly early in my military career, I was fortunate to get caught up in an enterprise called the National Reconnaissance Program (NRP) where no weapons were being readied but the commitment of those involved to do what needed to be done was unparalleled. And there was another element that bolstered this group's dedication to the country's national security; that of being in on a big secret, part of the in-crowd. While information is power, it is also a great motivator. This became very evident to me every time I was briefed into another Top Secret codeword project; at one time I was privy to over twenty of these compartmented programs and somewhat proud of it. So being in the know was a very powerful incentive to totally commit one's self to the task at hand.

The Sojourner

Being an unformed college graduate entering this environment, the first significant thing that affected me was the choice of a role model and I picked a strange one: General "Vinegar Joe" Stilwell. Generally aware of his WWII exploits in China/Burma/India theater, reading the *Stilwell Papers* focused my thinking on 'performance'; for years afterwards, I would regularly quote him as saying "I can only do my best and the hell with it." It turns out he probably never said those words. But on dealing with Chaing Kai-shek he wrote "My only concern is telling him the truth and getting about my business. If I can't get by that way, the hell with it." As for his imminent recall due to political considerations, he said "FDR proceeds to cut my throat and fire me. They can't hurt me. I've done my best and stood up for American interests. To hell with them." The bottom line for me was a mind set where performance was key and recognition for good performance only a secondary measure of success, i.e. good fitness (evaluation) reports, promotions, rank, etc. were not all that important.

Early on I also recognized that luck was going to be a factor in what happened. Getting a job ill suited to my limited skills or one I didn't enjoy was not something to be resented. Things just happened that way. And I was smart enough to understand that what turned out to be a bad break in one way, sometimes advanced or improved me in another. I like the way Kurt Vonnegut purportedly addressed MIT graduates on the subject (Mary Schmich actually wrote the words). "Whatever you do, don't congratulate yourself too much, or berate yourself either. Your choices are half chance. So are everybody else's." I also had a good grasp on how to handle mistakes. I tried to focus on conceptual understanding rather than details and thus was fairly successful in avoiding fundamental, gross errors.

As for the myriad little mistakes that inevitably crept into my work, I readily acknowledged their existence and worked hard to correct them. Another thing that helped is that I joined the Navy with very low expectations; it was very easy for me to enjoy even the most mundane activity. Every day was an adventure. A rewarding experience was not something I thought much about because it was not a part of my mental makeup to begin with. This attitude was clearly a hangover from my early participation in athletics, something that I fully enjoyed even though I can look back now knowing I mostly played on teams that were perennial losers.

Accompanying all of this was an ability to exhibit what it those days was known as a "gung ho" attitude. I don't think anyone I ever worked with thought that I would do anything less than try my best. Being at my desk 5:30 in morning and leaving at 8 in the evening was something I just did. In many ways, I was a counter example to the view 'an ounce of image is worth a pound of performance.' Getting the job done was the only thing that seemed to count in my work world. It was not until years later after reading *The Soul of a New Machine* that I realized what happened to me. Tracy Kidder wrote about the development of Data General's Eclipse computer this way: "There was, it appeared, a mysterious rite of initiation through which, in one way or another, almost every member of the team passed. The term that the old hands used for this rite ... was 'signing up.' By signing up for the project you agreed to do whatever was necessary for success. You agreed to forsake, if necessary, family, hobbies, and friends -- if you had any of these left (and you might not, if you had signed up too many times before). When you signed up, you in effect declared: "I want to do this job and I'll give it my heart and soul."

Fortunately, there were a number of kindred spirits that I worked for and with who felt the same way. Mission accomplishment was key to them also, and a byproduct of this was a concern about who their successor would be. Repeatedly, contemporaries would recommend me to replace them because they saw in me someone who could be counted on to continue carrying their torch. And the same thing happened with my bosses, who at every opportunity they could, advanced me to a more responsible job. As you will read, it was a fun filled 24 years with only a few hiccups.

In the years since I retired, I often wonder what it would have been like being a flag officer. I am positive now, as I was when I decided to retire, that selection for Admiral was not in the cards. One missing key element was my lack of either a professional or social network. Other than the standard cocktail parties, office picnics and golf outings, I never hobnobbed socially with my bosses or coworkers and I was remiss in keeping touch with most of my former shipmates, bosses and associates. And not having an entourage, if I had the opportunity to place 'my people' in critical slots, those key people just wouldn't have been there for me.

But more germane was my reputation for being 'difficult.' It seems that along the way I had fully embraced James Surowiecki's view that "the best collective decisions are the product of disagreement and contest, not consensus and compromise." As a result I am sure that no one ever came out of a committee meeting where something of import was at stake without muttering to himself, "what a son of a bitch that guy Frank is."

Charm was not something I really understood and certainly not something I thought all that important.

Below is my last official Bureau of Naval Personnel portrait photo from my service jacket. I like it since it hints of steeliness, something I thought I exhibited a lot of.

Finally, in advancing the merits of this memoir, I have some Theodore Roszak words for you to ponder. In his 1972 book about living *Where The Wasteland Ends*, he wrote: "There is nothing to do, nowhere to get. We need only stand still in the light."

Chapter 1- Naval Training (1955)

Officer Candidate School (OCS)

Right after graduating from college in May 1954, I enlisted in the United States Navy in the special rate of Officer Candidate; this enlistment would not take effect until November when my orders to active duty would be written. My reasons for choosing the Navy were varied but basically I wanted to escape my humdrum life for some adventure. I had not yet met my wife-to-be and if I had, I probably would have wound up doing something different. In any case, after making the decision, I spent most of the summer roaming around the east coast selling coupon books (dry cleaning, dairy products, gas station services) and then in the Fall got a job at Montgomery Ward selling men's clothes. Then having already rolled the dice, six months after enlisting, I journeyed to Newport RI and started OCS Class 20 on January 3, 1955. Life was simple but austere in OCS; the weather was windy and cold and the pay poor ($78 a month). But I did get to wear a neat uniform: pullover jersey top with bell-bottom trousers and a dixie cup, white hat (which I still have). Days were long; we got up at 6AM and normally packed it in at 11PM. It was a four-month ordeal no different than boot camp any other military recruit has to suffer through. The idea behind the whole thing was to test your ability to stand up to stress, mentally and physically. The curriculum was divided into six general subject areas: engineering, operations, seamanship, navigation, naval orientation and weapons. I did not have any trouble with the course work; all my grades were adequate, i.e., 3.0 or higher. I also survived the physical side of it in terms of drills, military bearing and decorum, etc. I think I did so well primarily not to let down my mother and aunts who would have been very disappointed if I didn't get a commission. I was the oldest living male in the family at the time, and the first one to get a college degree, so you can see how this might have influenced me. And I had good reason to be concerned about being booted out, since that is exactly what happened to two of my roommates (four of us shared a room). One was ill equipped mentally to handle the academic side but the other had a Master's Degree in Finance from the University of Chicago. His Achilles heel was a failure to cope with

the nonsense of the OCS environment; he couldn't keep his head above water, never seeming to accept the great truth taught at OCS: 'there's the right way, the wrong way and the navy way.'

But there was something far more important for me to accomplish than getting through OCS and that was pursuing the affections of a lovely 18-year girl named Mary Gallagher who lived in South Ozone Park NY. Her attractions lured me back to New York every chance I got. We had met at the sock counter in Montgomery Ward, Jamaica NY in early December, had two dates before I left for Newport, one a combined New Year's Eve/Bon Voyage party at my apartment where she acted as the hostess. I was absolutely crazy about her and regularly took the train from Providence RI to New York City those Friday nights when I got a weekend pass. On Saturdays we would do things like ice skate at Central Park or attend jazz concerts and wind up the day, sitting on the couch in her house 'talking.' Then Sunday afternoon I would board the train to get back to Newport. After four months of this sometimes grueling training, I graduated; both my mother and Mary trained up to attend the commissioning ceremony. The photo on the right, showing the three of us after the event, dates from May 1955.

Aviation Ground Officer School

After a week at home, I was off to the first bit of specialized training related to my aviation ground officer designator (1355). In mid-May, I flew to Jacksonville FL to attend the Aviation Ground Officer (AGO) school, and that event in itself was a real eye-opener. I had never flown before and I was really taken with the sensation of being pushed back into my seat as the aircraft took off. I said to myself then this looks like a lot of fun to do, and I will try more of it. The AGO course was eight weeks long and I remember next to nothing about it. I had my own room in the Bachelor Officer Quarters (BOQ) and aside from a short orientation cruise on the aircraft carrier Valley Forge and getting very drunk on Manhattans one terrible night, nothing struck me as momentous about my time there. When I finished up the training, I shared a car ride back to New York with some fellow graduates and spent time with Mary before I had to show up at my next duty station, one much closer to home.

Photo Interpretation Center

While I was at Newport, I received orders to report (after finishing AGO school) to the Naval Intelligence School in Anacostia, Washington DC for training in Air Intelligence. Just before I left OCS however, those orders were modified to report to the Photo Interpretation Center (PIC) also in Anacostia, for PI training. In late July 1955 then, I started in the Tactical Photo Interpretation Course, 12 weeks long, focused on terrain and military facilities. I was part of a fairly large class, many of whom were officers being retrained from some other specialty. I did fairly well but I never did become a good PI. There is some quality of being able to discern changes and see anomalies when looking at a photograph that I just did not have. I think it is a lot like being able to mentally work n-dimensional math problems. Again I have little recollection of what we studied with one exception. Part of the training involved interpreting land features, and a local area which we could ground check after looking at aerial photographs was Fort Washington NP. Located at a bend in the Potomac River south of DC on the Maryland side of the river, I can remember trooping up and down the hills, never suspecting that some 20 years later I would live just blocks away and watch my children run those same hills in cross-country competitions.

When I completed the tactical phase, I stayed on for Strategic Interpretation—ten weeks, smaller class size, emphasizing industries, transportation, etc. And while I may not have been a good photo interpreter, I began to tinker with photogrammetry, and got pretty adept at it. For fun and showoff, I measured object sizes on ground photographs, mostly snap shots, and I still have the Manual of Photogrammetry I used as a reference guide. It was nice to do something well like this and it helped me get ready for my first real Navy experience coming up shortly. My travels up to New York every weekend continuing my romance with Mary became routine, (photo at right was taken in the late summer 1955) and to make this easier, I bought a car, a 1949 Olds 98. To afford it, I got a night job with Griffith Consumer supervising a crew taking phone orders for home delivery of heating oil. I eventually got fired from the job for 'dissing' a fellow nighttime supervisor, one who had a lot of seniority. Unfortunately it would not be the last time this kind of

thing happened. I graduated in early January 1957 with a set of military orders to 'go west.' During this time, Mary and I talked about getting married but given the uncertainty of what was looming ahead, discussion was as far as it went. But I don't think either one of us thought it wouldn't happen, sooner or later.

Chapter 2–Pacific Fleet (1956-1959)

Composite Squadron Sixty One (VC-61)

My first operational assignment was to VC-61, a composite squadron based at NAS Miramar located just north of San Diego CA. I took off from NY in early January heading south in my Olds, stopping the first night in a blinding snowstorm outside Charlotte NC. After that it was smooth sailing with overnight stops in BOQs at Maxwell AFB in Montgomery AL, NAS Dallas, Biggs AFB in El Paso, and NAS El Centro in the California desert. Miramar had just re-opened after being shut down for ten years, and as the new BOQ was full, I took up residence in an old WWII barracks (see photo on right).

Composite squadrons like VC-61 were being phased out; it would be re-designated a photo squadron (VFP-61) later that year but it still had a variety of aircraft to include twin engine Beechcrafts and TV-2s, (the Navy equivalent of the Air Force Shooting Star) and one F9F-8T (trainer version of the then, first line Grumman fighter nicknamed the Cougar). I was appointed the Classified Material Control Officer and Atomic Biological Chemical Defense (ABCD) Officer, both essentially non-jobs, which freed up a lot of time and allowed me to learn to co-pilot the Beechcraft and mooch hops in the backseat of the TV-2s and 8T. Before I could do that however, I had to first experience hypoxia in a low-pressure chamber and simulate a cockpit ejection on a test sled. The oxygen deprivation drill was an informative. Teamed with a Chief Petty Officer, after removing our oxygen masks, we were directed to play patty-cake. I was amazed at how soon he lost awareness of what was going on, failing to respond to our instructor's shout to "Put your mask back on!" The sled test was a gut check to see if you could pull the curtain and the only thing I can remember about the experience was coming down without any memory about the initial kick in the ass or the stop at the top. On the right is my 'OMIAS' card which commemorates the event.

NAS Miramar, 13 miles north of San Diego, was back in the mid 50s isolated and easy to get to unless there was fog. One night after a visit to town, I had to navigate the freeway by looking down at the left highway stripe from an open car door to see where the line ended and the left turn to base gate began. The photo below shows the base then with a flight of Chance Vought F7U Cutlasses preparing to land.

To give you some sense of what it was like being in a squadron, excerpted below is a portion of the April 1956 squadron newsletter about a cross-country trip I was involved in. Excerpts from other squadron newsletters are in Appendix B.

> A cross-country flight often is one of the most uneventful experiences but every once in a while you get one in which everything goes haywire. John Romano and Len Derse in 5-Ps were to accompany Ralph Sykes in the TV-2 to drop off Walt Zimbeck in Atlantic City last weekend. Eastbound 80 miles from Tulsa, Len flamed out. After a couple of failed restarts, at 10,000 feet with one hand on the face curtain, Len spotted a good sized airfield and dropped down on the runway, just over the approach end. It was Tulsa airfield, 80 miles dead stick. Len stayed with his plane while the other two aircraft continued eastward. But the perils were just beginning. Ralph picked up Frank Bausch in Atlantic City who had spent the Easter weekend with his girlfriend in NYC. On the way back, Ralph lost his radio after leaving Omaha and John experienced auto acceleration at the same time. Finally the 5P flamed out and down went John. A shotgun start bore fruit and John was denied the joys of ejecting in the Rocky Rockies. Ralph with no radio thought John was disgusted with his sloppy lead and was descending to a more favorable altitude. In the meantime Frank watched all this in the rear seat with a slashed radio cord as a result of Ralph using his penknife as a helmet stretcher. He was the only one who didn't panic. He thought they were just being playful with all the hand waving.

This same Ralph Sykes and I had another adventure on a trip north to the Bay Area (NAS Alameda) in a photo Beechcraft where I was in the right seat. Immediately after we passed over LA, we ran a fuel tank dry and the silence was frightening before Ralph switched tanks and hit the wobble pump to get some fuel back into the engine. I took the photo on the right of a TV-2 from the backseat of another one. Essentially what I would do to get all these rides was to bribe the pilots by offering to fly their OFT (Operational Flight Trainer) hops, which they abhorred but I loved. I would sign in using their names; the techs running the equipment didn't care all that much and helped me learn to fly the trainer without destroying it (I crashed a lot especially during takeoffs and landings).

One actual hop was especially memorable as pilot Jerry Olson went through the loop, spin and roll routine trying to see whether he could get me to barf in my oxygen mask. When not in the air, I was the squadron go-fer, haunted the ready room listening to mission briefings and debriefings, went to sea with every detachment which was carrier qualifying, hung out on the maintenance line or in the photo lab, and made myself otherwise useful by setting up the 16mm movie projector for the *Victory at Sea* episodes everyone watched at lunchtime. Two shipmates made a big impression on me: Ben Cloud, a Native American and Reeves "Rip" Taylor, a Black American. Both were top-notch professionals and set examples for their squadron mates with their performance. And we had an RAF exchange pilot, Captain "Fearless" Fosdick who took a merciless ribbing with good nature as Al Capp's 'Lil Abner' was popular at the time. Overall a great squadron and a wonderful introduction to naval aviation.

Another maturing experience unrelated to flying was an investigation I conducted that involved a Chief Petty Officer who had the left the Chief's Club late at night and drove the wrong way onto the freeway, killing himself and a driver in another car he hit head-on. I interviewed probably about ten of his compatriots at the bar that night as well as the bar tender on duty, and all I could get out of them was "the chief had a few beers" (I don't' remember a blood test being done). Then I was both angry and surprised at what I knew was a cover-up but now I better understand why their concern shifted to damage control. They were worried about the potential impact on the chief's family image-wise, and in respect to their getting survivor benefits.

WestPac Cruise

My first Western Pacific deployment began in May 1956 along with four other VC-61 officers (Verg Brown, Tom Mauney, Don Lytle and Slim Tinsley) aboard the USS Lexington (CVA-16) with a three plane F9F-8P detachment. The Lex was a WWII vintage carrier originally commissioned in 1943 and home-ported at North Island, a short ferry ride from the then sleepy town of San Diego. I spent the next seven months aboard her in what is called the 'Junior Officer Bunkroom' located in the bow of the ship just below the flight deck. The Lexington is still afloat today, serving as a navy museum in Corpus Christi TX (see photo below).

Our detachment designator was "Hotel" and we operated out of the same ready room as the VF-111 Sundowners (they flew the fighter version of the Cougar). The photo below, taken with Mt. Fuji in Japan as a backdrop shows a Sundowner flight of F9F-8s.

While our air group, ATG-1, had a standard mix of squadrons, it deployed with a varied assortment of detachments. Besides the usual Photo, Airborne Early Warning and Helo guys, we had two pilots from a Guided Missile Group (GMGRU-1) flying Cougars which controlled Regulus missiles via radio control, flying wing on them during their terminal dive; Heavy Attack Squadron-6 with AJ-2s bombers which took up so much deck space as to make routine flight operations bizarre; and a VX-4 group experimenting with Sparrow missile equipped F7U Cutlasses. With this exotic blend of aircraft, everyone with spare time spent it watching launch and recovery operations. And for those who have never seen them, it is one of the greatest shows on earth. The image below is the cover of our end of cruise report with a sequence of forward firing photos showing a carrier landing, a favorite giveaway from our detachment to other squadrons.

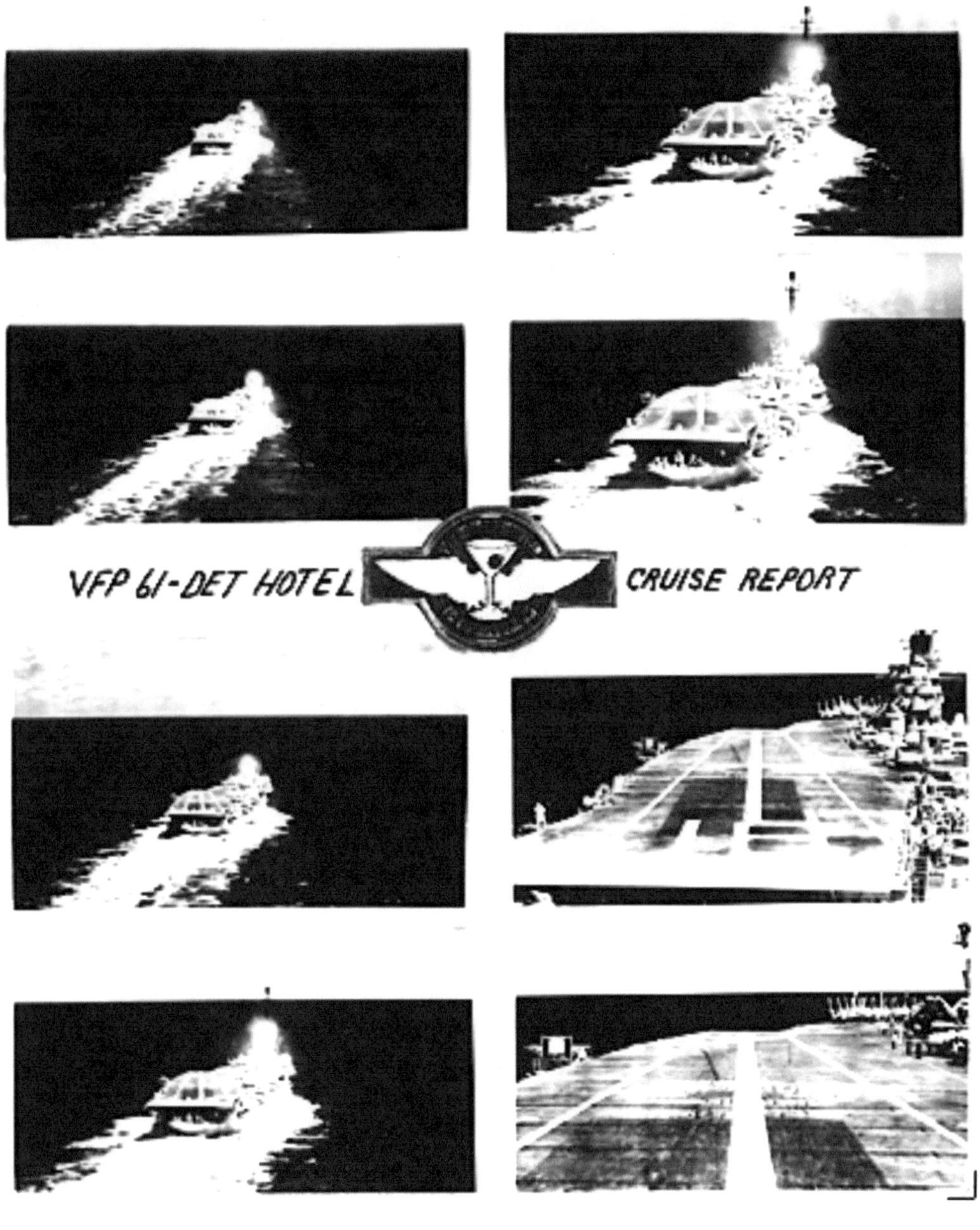

The centerpiece of this montage is a reproduction of a patch the detachment wore on their flight jackets during the deployment. In my pile of memorabilia, I have the original, which features a martini glass (with olive) in the center of the navy wings.

The photo on the right was taken of me in dress whites aboard the Lex after a parade formation on the flight deck; one of our photo birds serves as a backdrop. The cruise also presented to me for the first time, a variety of sight-seeing opportunities. The ship visited numerous ports in Japan (Nara, Yokohama, Kobe, Sasebo) Subic Bay in the Philippine Islands, and Hong Kong; the excursion I most enjoyed was in Japan to climb Mt. Fuji (12,388'). Along with shipmate Don Lytle and his sister, we traveled from Tokyo west by train and then used a tram to get to the Fuji View Hotel. The next day we traveled by horseback to the base of the mountain and then overnight, climbed to the top so as to be present to see the 'rising sun.' The contemporary map below shows the relationship of Fuji to its environs.

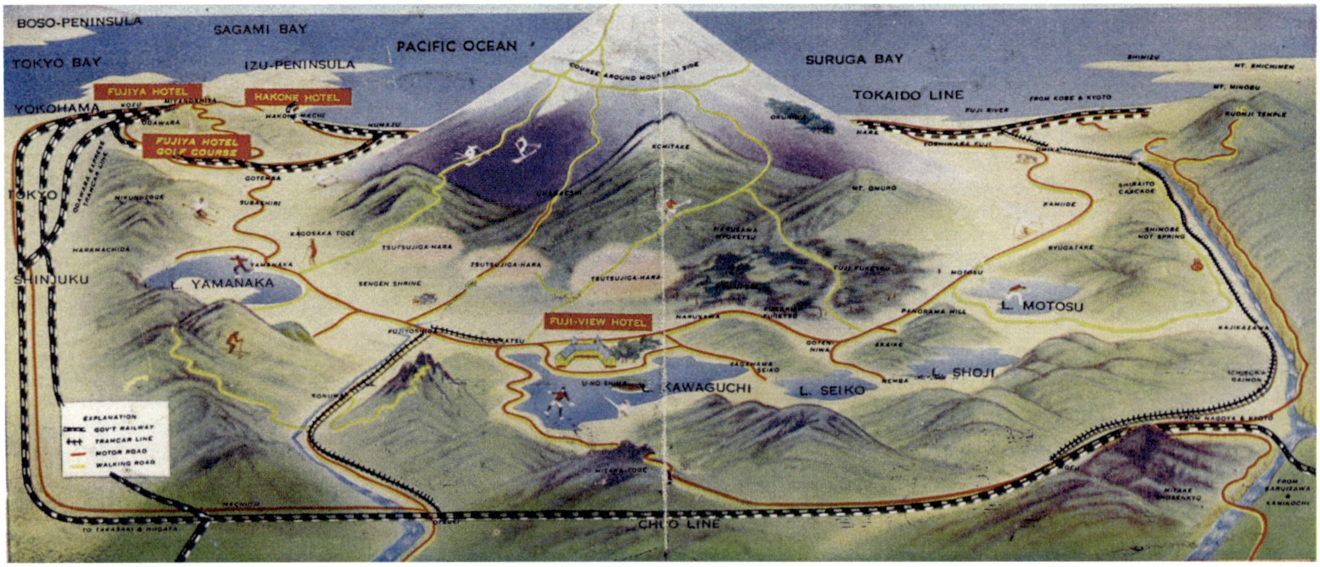

I still have my Fuji walking stick that was branded at every station on the mountain where vendors provided hot white tea. It was a simple climb which featured repeated hair-pin turns. The photo at the top of the next page was taken at dawn from the crater's edge and may explain why being there at dawn was worth every ounce of effort to get to the top. It is a pilgrimage for many Japanese and their chanting as the sun broke over the horizon sent chills down my spine. The descent was another plus as we took large giant steps on a lava flow field, which amounted to one great slide downhill. But the venture that

made the biggest impact on me was not near as enjoyable as the mountain climb but more memorable; it was to Hiroshima. Eleven years after the dropping of the bomb, the city had not been rebuilt and pretty much looked like it had been bombed the day before. I took a lot of photos but they do not adequately give one the sense of the desolation and devastation caused by the nuclear attack that ended WWII.

Even when in port, the Lexington operated aircraft. In September 1956, we launched most of the air group tied up to a pier in Yokosuka. The photo on the right shows one of our Banshee night fighters being catapulted off the flight deck in the harbor. During flight ops, I spent a lot of time on deck readying our planes for photo training hops or in various squadron ready rooms just listening to briefings. We also had on board a number of veteran aviators, the most notable being

Warren "Andy" Skon, Operations Officer for ATG-1, a WW II Ace, and Butch O'Hare's wingman.

A photo of our 'Cruise Board,' illustrating the deployment highlights is on the following page. Upon our return, the 2'x3' board, done in full color, was hung in the squadron ready room at Miramar. Unfortunately my reproduction is in B&W and somewhat blurred. You sort of have to take my word for it that the original was beautiful.

The detachment had only one real mission, about which I have been unable to find a declassified reference. Our detachment leader, LCDR Verg Brown, had been awarded a Distinguished Flying Cross for his exploits on his previous deployment, and didn't bat an eye about conducting such flights. I was a spectator during the planning phase for this operation, which was overseen by the CarDiv-1 intelligence staff. After mission launch, I monitored the event in Combat, saying silently to myself "Are we really doing this? " and then aloud "Lord, No!" as one of our guys broke radio silence with "Bogey at..." resulting in a late-mission abort as our planes dove for the deck to get back home safely. My only other contact with the intelligence world these seven months was ENS. J. T. Miller, VF-111's AI (Air Intelligence Officer). Anyone who knew Ted as an irrepressible CDR, can only imagine what he was like as a fledgling AI in a fighter squadron.

Over the years, I would be involved in a number of 'over-flight' missions but none ever seemed to me to be as exciting as missions launched off a carrier using aircraft not designed for that purpose. Two prime examples are the U-2 and the C-130 Hercules. Neither could initiate their mission satisfactorily from any land base, and while they could not be catapulted, they could be deck launched into the wind, becoming airborne as they crossed the bow, an evolution requiring a lot of wind and ship speed; they would eventually land at some airfield within range. One over-flight I became involved with happened years later, when a C-130 mission dropped remote sensors deep inside a country with a sophisticated air defense system, doing so by flying as low as possible with multiple jinks on its flight path. Here again, I could hardly believe we were doing it but that was the nature of the business I had gotten into.

Shore Based Respite

I proposed to Mary by mail sending along a diamond engagement ring and she accepted. My financial situation had improved with my promotion to LTJG in November and when I got back to the US just before Christmas, I flew to NYC and on December 29, 1956 we were married in her parish church, St. Clement Pope. Friends and family attended, and after the wedding reception, we took off in her car to return to San Diego. The drive back to Miramar served as our honeymoon as we spent three glorious weeks wending our way cross-country. We drove south to New Orleans and a spent a lot of time in the southwest stopping at Carlsbad Caverns, Grand Canyon, Bryce and Zion National Parks, finally winding up in Las Vegas before it all ended. I reported back to the squadron, and after a short stay in base housing (a Quonset Hut located on the east end of the main runway), we found a lovely apartment in Del Mar just two blocks from the ocean. We began socializing with a number of young married couples in the squadron, started playing bridge and I had my first martini. We mostly hung out on the beach, with me doing some bodysurfing. One sightseeing trip was to Disneyland with Jack and Althea Schroeder, where the photo on the right was taken.

My romance with naval aviation continued as I broke the sound barrier. Don Lytle, who I had climbed Mt. Fuji with, took me up over Camp Pendleton. To go supersonic in a F9F-8T in hot weather, you had to climb for almost an hour to reach max altitude (35-40,000'), then flip over on your back and head straight down. What a kick, especially experiencing the G forces pulling up. Those first seven months of 1957 were so much fun that they couldn't possibly last and sure enough they ended abruptly with an unexpected transfer in July to the staff of Commander Carrier Division Five. My squadron sendoff poster is shown on the next page.

Carrier Division Five

I had little time left in San Diego, as on 16 September I left aboard the USS Ticonderoga (CVA-14) on my second WestPac cruise. Mary, now very pregnant, went back to New York to stay with her parents and our son Rich was born there on 19 September, just as the Ticonderoga got to Hawaii (notification message below).

ACTION
COMCARDIV 5

WU KC
CW ...EMAN
ORIG OR ACTION LTJG BAUSCH
METHOD XMIT HR 3498

UNCLASSIFIED DEFERRED

MSG JAMAICA NY SEP 19 LTJG FRANCIS BAUSCH RICHIE BAUSCH WEIGHING IN AT 8-12 ARRIVED 1700 THRUSDAY THE FATTEST ROSIEST ROSIEST BABY IN THE NURSERY LOVE MOM

Making the best of a tough situation, I directed my efforts towards qualifying as a Staff Tactical Watch Officer which among other things, entailed standing watch on the flag bridge, sequencing signal flags necessary to turn/corpen the task group (usually one carrier, four destroyers and a cruiser) into the wind for a launch. In this evolution I was especially mindful of the Columbus/Parks collision that occurred almost two years earlier as a result of a foul-up by a number of entities including the Carrier Division Five staff. An eight ship diamond formation (two carriers, two cruisers and four destroyers) were operating at night using visual signaling (flashing light) and directed to 'Turn 035, Speed 20' at 0400 local time. The signal was executed only by the Columbus which almost sheared off the bow of the Parks. The photo below shows the damage inflicted on the destroyer which resulted in a two fatalities.

So I would carefully check and recheck the flag hoist with our signalmen, go over the whole thing with our ship's Officer of the Deck (OOD), and then with trepidation as the Officer in Tactical Command (OTC), execute the maneuver and watch as everyone moved in the right direction at the same time, stopping at the proper heading. Even today, it still strikes me as amazing how much trust and confidence the Navy, in the persons of RADM Fitzhugh Lee, CAPT William. I. Martin and Surface Ops Officer CDR Will Trice put into a 24 old year novice. Not only was this particular experience a maturing one, it made me think for the first time of making the Navy a career.

There was one maneuver however where I was not qualified to act as OTC and that was a replenishment at sea. Keeping a navy task group stocked with adequate supplies is generally done while underway, and coming alongside tankers, stores, reefer, and ammo ships in formation is a very dangerous thing to do. The photos below show, on the left, an oil painting by Walter Brightwell of a destroyer being topped off with fuel from an aircraft carrier (this photo was on the cover of the December 1970 Naval Institute Proceedings). The photo on the right was taken by friend Ron Jones of a routine replenishment of a task group; here the oiler Chuckawan (AO-100) is simultaneously refueling the Guided Missile Destroyer John King (DDG-3) and the carrier USS Forrestal (CVA-59). You can see from the wealth of white water and the proximity of ships to each other, this is not a maneuver you want to have a rookie oversee the action—as the saying goes 'a collision at sea can ruin your whole day.'

We did one replenishment on a dark winter's night in the Sea of Japan off Hokkaido with all the ships in complete radio/radar silence, being directed to the rendezvous by one-way radio from an airborne controller. Posted on the flag bridge was a sign which read "*If you can keep your wits about you while all others are losing theirs, then you evidently don't understand the situation.*" During this maneuver, there was no confusion on the bridge, just grim silence waiting for a visual confirmation that things were going right. The pressure on the Admiral, who was genuinely worried about a potential disaster, got so intense, he had us turn on our radar just before joining up.

Concerning my responsibilities as the staff Photo Interpretation Officer, they were virtually nil. Only one assignment, a survey of Far East USAF base loading, made use of my skills. I mostly worked on various other projects assigned by the staff intelligence officer, CDR Wilbur C. Griese, known to his peers as 'Wil.' He had been awarded a Navy Cross for a July 1945 bombing mission that attacked the Japanese battleship Hyuga at Japan's Kure navy base. Coincidentally he had flown his TBM from the same carrier we were presently on, the Ticonderoga. He had been called back to active duty for the Korean conflict where he was awarded a Distinguished Flying Cross. But being a reservist, he was very worried about whether he would be a victim of a reduction in force (RIF) as he married late and had six small children. He eventually made Captain retiring after 22 years of credited service, went to work in the general aviation field in St. Louis and lived to be 89. But his situation then raised a big caution flag with me about trying to make the navy a career.

We changed our flagship in November and for the rest of the cruise were embarked on the USS Kearsarge (CVA-33); the photo on the right was taken on its flight deck. Its Air Group was much less exotic than e.g., the Lexington's, but its veteran leadership made up for it. CDR Paul Gray, nicknamed the 'Butcher of Kapsan' for a notable Korean War air strike, was the Air Group Commander and Kenny Shugart, his wingman on that mission, was the Operations Officer. As for my peers, there were two other 1355 designated air intelligence officers on the staff, both LTJGs like me, but I was not impressed with either of them, mostly because they did not seem all that interested or excited in what was going on around us. And my roommate was the ship's lawyer who was a grade 'A' malcontent. So for a lot of reasons other than missing my wife very much, it was not all that fun of a cruise.

Port calls and liberty excursions were generally to the same places the Lexington had been the year before. When the ship anchored in Subic Bay in the Philippines, I got to visit Olongapo for the first time, a place more of a cesspool than Tijuana. The visit that I remember most fondly was the return to Hong Kong. Then a British Crown Colony, the island is nestled on the coast adjoining Communist China, which obviously prohibited excursions of US sailors into the country-side, unlike Japan where you could travel anywhere. But there was much to see in the colony itself: a trip by tram to the top of Victoria Peak, a walk through town past the Bank of China or to the docks to see the boat people, a ferry ride across the harbor to spend time in the lobby of the Peninsular Hotel and see the Kowloon railroad station where you might board a train to Canton. Flying into Kai Tak Airport (now closed) was also a memorable experience as the landing to Runway 13 required a sharp right bank or else crash into the mountains immediately northeast of the airport. But the event I remember most clearly is the ship's approach to our anchorage in Victoria Harbor. At 0745 that morning, I assumed the watch while we were still in the South China Sea and when I reached the bridge, to my amazement in front of us spread out across the horizon directly in our path were a mass of sampans/junks like the one shown on the next page.

I called the ship's OOD, the OTC at the time, and asked him what he was going to do. A veteran of this type meeting, he said: "nothing-we'll just maintain course and speed and they'll get out of our way." And sure enough, the fishing fleet magically parted in minutes to let us pass, just like the Red Sea did for Moses. I'm not sure what the Rules of the Road required but the OOD's approach sure worked.

We returned to North Island in April 1958; reunited with my wife, for the first time I saw my now seven-month old son, Richard. My mother also flew out to become part of the family, either living with or near us, for the next 30 years. There were some anxious moments when I soon found myself in limbo, i.e., fairly certain I was getting off the staff but not knowing when, or where I would next be assigned. But the AirPac Intelligence officer got on the phone to Washington and in May 1958 I received orders to the Pacific Fleet Air Intelligence Training Center (PFAITC) at NAS Alameda CA.

Pacific Fleet Air Intelligence Training Center

NAS Alameda was located in the San Francisco Bay Area just south of Oakland and I would spend almost two years working there. We bought a 1951 Lincoln Cosmopolitan in San Diego, used it to haul a small trailer with all our belongings to our new home, a three bedroom rental in San Leandro just south of Alameda. I was assigned as an administrator/instructor in photo interpretation, which required the use of special instructional materials. To give one example, I would provide students with stereo photographs of the industrial area shown on the next page and ask them to identify the circled areas or determine things like the speed of the boat at "FF" (top of photo).

The skipper of PFAITC was CDR Jack Newby, a classmate of mine in Tactical PI School back in 1955, who lived just south of us in San Lorenzo. He drove me to work pretty much every day and we got along fine if you ignore the volleyball smash I delivered during a lunch-time game that nearly broke his nose. I don't think he ever really forgave me for that. For most of my time at the center, my title was the Photo Interpretation Training Officer, differentiated from my Air Intelligence counterpart, who oversaw a curriculum quite different from mine. As an instructor, I did not do very well; the evaluations provided by the students at the end of the course indicated I was a pretty mediocre teacher. I did however take the AI course, learned a little bit and banked my reservations about the efficacy of AI/PI training in a mental vault which I would open up 12 years later in 1971 when I became the Deputy Director of the Armed Forces Air Intelligence Training Center in Denver.

Weapons Employment/Target Folders/Recreation

One of the benefits of my PI training was a fairly good understanding of targets and I soon put that to good use. I was moved out of the training slot in early 1959 and took over as Personnel Officer. I had little interest in this assignment and while competent in administrative affairs; it still was a terrible mismatch of man to job. But it did give me some time to address a new problem that arose. The Navy's weapons employment manual, NWIP 20-1, had become seriously out of date, and the Center was directed to begin training air intelligence officers in conventional weapons targeting. So part-timing it as the Personnel Officer, I developed an employment manual with lots of target illustrations along with their functional/structural vulnerabilities, sort of a homegrown Joint Munitions Effectiveness Manual. I made copies for some of the deploying shipboard AIs and felt good about the project.

While that was sort of an unheralded success, I also had a very visible setback with a fairly straightforward task assigned me, that of developing training target folders similar to those produced on an aircraft carrier. I misinterpreted my guidance, in that I produced some pretty shoddy but realistic looking stuff which one could envision coming out of a no-frills field operation. But as I subsequently found out, the real, but unstated requirement was for slick, nice looking production material. As a result, I got replaced after a couple of months on the project and the new Executive Officer, CDR Bob Burroughs, who had experience doing this kind of work at the Fleet Intelligence Center Pacific took over the task.

With not a lot of job pressure, it is surprising how little Mary and I got to reconnoiter the area. We had two family excursions: a trip to Yosemite NP with Grandpa and Grandma Gallagher, made very memorable by Rich getting terribly carsick, and a overnight trip to Clear Lake towing a camper trailer where the Lincoln experienced a fan belt failure. We got to the beach once, where unfortunately the water was so cold that swimming was out of the question and we never tried that again. But we did go into San Francisco every so often, most notably to eat at the Iron Pot where all customers when seated, were served a monstrous amount of minestrone in—an iron pot.

The PFAITC staff was socially diverse. Larry Keenan, brother-in-law of future presidential candidate Mitt Romney was somewhat aloof but another, Al Ragan, was a native of the Bay Area and got the instructors out to a San Francisco Giants game at the old San Francisco Seals stadium. It was here on Al's recommendation, I first ate peanuts in the shell. It was not so bad an as you may think, but you had to drink a lot of beer to make the experience palatable. Feeling comfortable in the area, when we had visitors, we would venture over to the Fisherman's Wharf tourist trap; I still have the DiMaggio's Restaurant menu from the time my cousin Eileen came to see us.

Getting Ahead and Moving On

I was selected for promotion to LT effective April 1959. I learned of the board action that previous December but in order to actually get promoted, I either had to take an examination or complete 18 prescribed correspondences courses, seven of which I had not even started. Not wanting to risk the exam, I choose the latter course of action, ordered the materials and that Christmas holiday period, Mary and I sat down and finished in the space of two weeks: Shipboard Electronics, General Aerology, Jet Engines, Air Plane Power Plants, Naval Airborne Ordnance, Aircraft Electrical Systems, and Aviation Operations. No wonder she is so savvy on all these subjects.

When the promotion finally came through, I remember one of the enlisted guys who worked for me, Bud Lovett, confiding that he was pleasantly 'surprised' by it. Now talk about a mixed message. Right before I left Alameda for my next duty station I applied for a regular commission as an Aviation Ground Officer (1350 designation). And then in early February 1960, with orders to shore duty in hand, I bundled the wife and kids up (we now had two-daughter Ann arrived in June 1959) and along with my mother, drove back east back into the ice and snow of Washington DC and the next phase of my navy career.

Chapter 3–Navy PIC (1960-1963)

In January 1960, I reported to the Naval Photographic Interpretation Center (NAVPIC) in Suitland MD, where all the Center's elements, other than training department, had recently moved from Anacostia. For the first time, we bought a house, a three-bedroom brick rambler in North Forestville. Hoping to make some money in selling the property, I began doing the 'home improvement' thing, and developed a penchant for it, evidenced with every house we bought thereafter. I am not sure I ever arrived at the break even point where the money spent exceeded the money earned but I did learn a lot about construction techniques. Another bonus about the house was its closeness to center which was located on Suitland Road in the government complex shared with the Census Bureau and the Naval Oceanographic Office. It would be the easiest commute I would ever have.

One of the oddities about NAVPIC was its command structure. Many technical centers with military personnel assigned have adopted the same approach to management but it seemed to me at the time to be a strange way to do things. All the departments had a military officer in charge with a civilian as a deputy. The Commanding Officer even had a civilian Technical Director backing him up. The four major Departments (Production, Special Projects, Evaluation and Technical Services), were headed by Navy LCDRs with GS-13 Assistant Department Heads. The idea was for the military officers to come in every few years with new ideas and a current understanding of what the navy needed done at NAVPIC. When they rotated out, the civilians, who pretty much were lifers in the job, provided technical expertise/continuity/corporate memory for the department. I was there three years and it worked fairly well. But I did not fully understand how the civilian assistant department heads looked at the whole thing and in mid-tour I provided advice to one that had unforeseen consequences for his career. More about this later.

Technical Services Department

I was initially assigned to the Technical Services Department as a Liaison Officer with primary duties to maintain a close working relationship with the CIA Graphics and Document Registries, both located on the mall in downtown Washington DC. I was usually given a name for some area in the Soviet Union and asked to obtain all existing intelligence material about it. For the first few months I was in the dark about why I was doing this. It was not until I was ushered down to the Special Projects Department conference room, signed an oath, and LT Jim Bartholomew briefed me on the then still covert U-2 over-flight program that I fully understood what was happening at NAVPIC. The information I had been gathering was being used to help hide the fact that U-2 imagery was the key ingredient of the intelligence flowing out of the center. And I was staggered as I looked at the briefing charts showing at that time, 20+ mission tracks flown over the Soviet Union since 1956. And this is how I was introduced to spook world of the National Reconnaissance Program (NRP).

In 1960, Corona satellites took over the brunt of imagery collection of the Soviet Union, with the U-2 and follow-on, more advanced photo recce aircraft used for special missions. During the next 30 years I would be involved in a wide variety of NRP activities, always marveling at the tremendous advances in imaging and data processing technology this program gave birth to. Readers used to Google satellite photos, digital cameras and Stuffit/Unstuffit would be surprised at how rudimentary things were back then.

In respect to satellite imagery systems, the diagram below shows the ground coverage pattern of the mature version of the Corona imagery satellite (KH-4) film return system (capsule ejected after reentry) with resolution on the order of 6-10'.

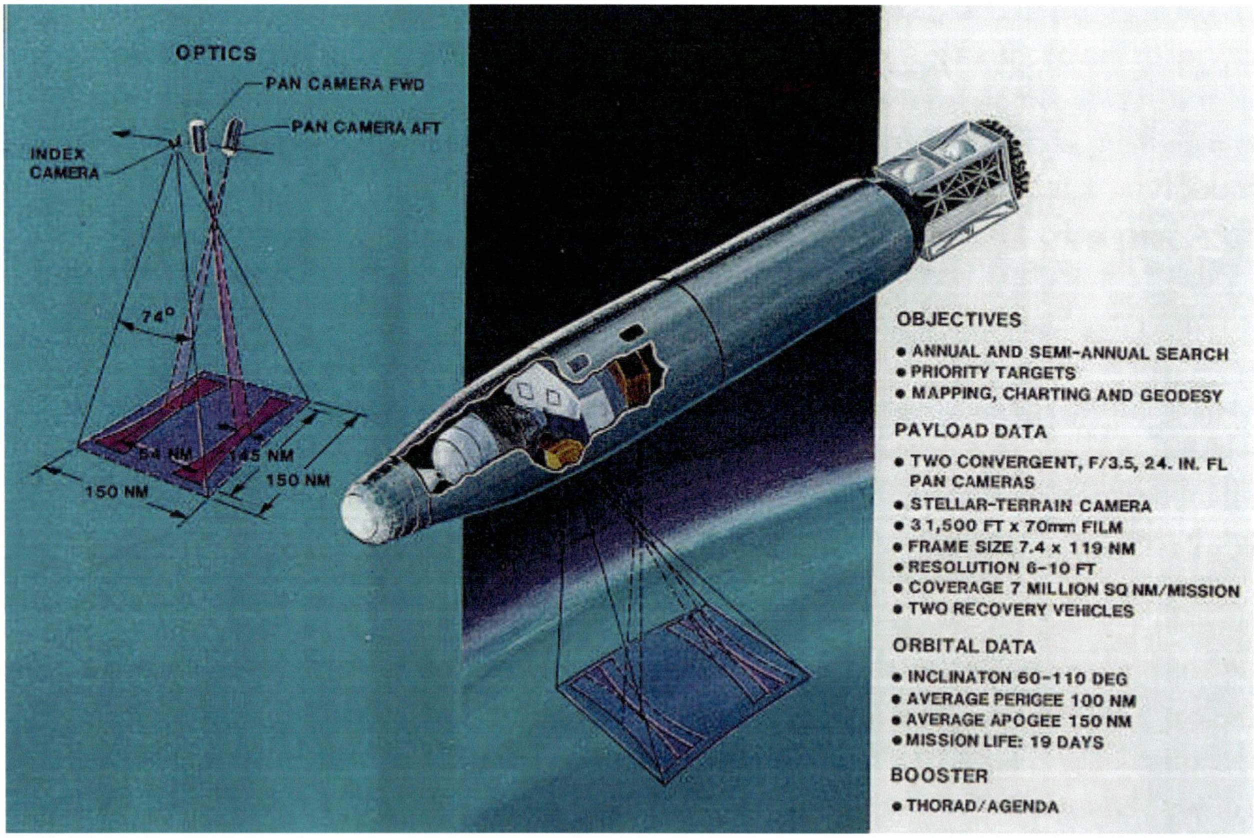

It is hard to imagine today but a key hurdle to overcome before the project became a success was the recovery of the film capsule. Using a net deployed from a C-119 Flying Boxcar, it would actually be caught mid-air over the ocean during its descent. As backup, a shipborne team would be deployed in the area for water recovery in case the airborne catch failed. In its later stages, the program shone; from 1966 to 1971, there were 32 consecutive successful launch-and-film-recoveries. But what was really amazing about this program was the persistence shown in keeping it alive in the early going. Below I have listed the first fourteen 1959/60 Discoverer launches (unclassified name for the initial Corona vehicles) and the results of each. It would be inconceivable to think that a program with such a dismal early track record could survive today. Curtiss Peebles has written a book on all of this, *The Corona Project*.

Early Discoverer (Corona) Satellite Launch History

Discoverer 1–2/28/59
1st polar orbiting satellite; KH-1 prototype; did not carry camera or film capsule
Discoverer 2–4/13/59
KH-1 prototype; tested capsule recovery techniques; did not carry camera; capsule recovery failed
Discoverer 3–6/3/59
KH-1 prototype; did not carry camera; film capsule recovery failed
Discoverer 4–6/25/59
KH-1; 1st generation low resolution photo surveillance, failed to achieve orbit.
Discoverer 5–8/13/59
KH-1; 1st generation low resolution photo surveillance; film capsule boosted into higher orbit, decayed 2/11/61
Discoverer 6–8/19/59
KH-1; 1st generation low resolution photo surveillance; film capsule recovery failed
Discoverer 7–11/7/59
KH-1; 1st generation low resolution photo surveillance; satellite tumbled; film capsule not recovered
Discoverer 8–11/20/59
KH-1; 1st generation low resolution photo surveillance; film capsule recovery failed
Discoverer 9–2/4/60
KH-1; 1st generation low resolution photo surveillance; premature 1st stage cutoff
Discoverer 10–2/19/60
KH-1; destroyed by range safety
Discoverer 11–4/15/60
KH-1; film capsule recovery failed
Discoverer 12–6/29/60
KH-1 prototype; designed to test capsule recovery system; did not carry camera
Discoverer 13–8/10/60
KH-1 prototype; designed to test capsule recovery system; did not carry camera; capsule successfully recovered from ocean
Discoverer 14–8/18/60
KH-1; film capsule recovered 1.2 days later; 1st successful photo surveillance mission

Division Officer

One of the less understood aspects of being in the military are the job responsibilities related to leadership, something which do not go away in a NAVPIC type environment. I was assigned as the Division Officer for all the navy enlisted men in the Technical Services Department (less Photo Lab) and I instituted both a reading and professional ethics program for them. I remember one session where we talked about *The Catcher in the Rye* and how one Chief Petty Officer was offended at the book's language. How quaint! An outgrowth of this effort was the establishment of a non-technical reading section in the Technical Library where anyone could check out paperback novels, non-fiction best sellers, etc. I weaseled money out of the Welfare and Recreation fund to buy the first couple of hundred titles and chose exactly which ones would be purchased. Never again would I have the feeling of wielding so much power.

Special Projects Department

In August 1960, the same month as the first successful Corona mission, I accepted a regular aviation ground officer (1350) commission. The following year, I was reassigned to the Special Projects Department and given the task of applying ADP techniques to the production of imagery intelligence reports. I continued to be involved in hiding the real source of the sensitive classified information contained in the Air Targeting Materials Program (ATMP). When the US started collecting satellite imagery classified Top Secret (Special Access Required), it began incorporating derived information from this imagery onto Secret ATMP charts using symbols, drawings, etc., and doing this very carefully. In general we needed to attribute a source other than imagery so as to keep the charts at the Secret classification level. To cite an example, the reference photos for the ATMP chart of the Soviet airfield on the right would likely have been German tactical recce imagery dating back to the 1940s. To update the chart to show the longer, modern runway and taxiway, the expanded parking aprons and the additions to maintenance and weapons storage facilities, we were required to cite sources other than recent photos, such as clandestine CIA reports.

I fabricated such reports at my desk in Suitland MD, sent them over to the CIA who introduced them into their reporting system at whatever point on the globe it made sense to do so. One of the reports that I remember writing had an itinerant construction worker flying from Site A to Site B in the Soviet Far East; he reported that he just happened to look out the plane's window and saw a big airfield whose runway had different shadings. He guessed that the more lightly shaded (newer concrete) at one end was about half again as long as the older surface. In this way the runway at the Arctic staging base, Mys Schmidta, 400 miles down the coast from Anadyr, got to be shown as 6000' long on the ATMP chart rather than previously documented 4000'.

And while my tour at NAVPIC gave me a new appreciation of the importance of national reconnaissance, it also changed my life in another very significant way: I started to play golf. My boss in Special Projects, LCDR Jim Mickelsen, saw some potential in me and before long I was spending every spare minute out on the course or the driving range. I became a practice nut and some evenings I would drive the family out to a vacant field on Andrews AFB and hit balls, which Mary and the kids shagged for me. The arrival of Jim (shown in Mary's arms in the June 1961 photo on the right), and John a year and a half later, filled our small house to max capacity but overall didn't change a lot of things other than make me much more aware of my family obligations. Job wise, I became totally involved in the Department's functioning, as I believed Special Projects did the most important work in the world. And as a result, based on my

advice and encouragement, the Assistant Department Head who I had worked for in Technical Services came to work in Special Projects as one of the Division Chiefs. Unfortunately, it was a terrible fit, primarily because I had fundamentally misjudged both his desire (which was nil) to move up the civilian work force ladder, something which had a pretty moribund advancement structure to start with. In effect I had essentially advised him to take a lower position in respect to his contemporaries. This episode pretty much soured me from that moment on about giving professional advice to anyone on how to advance. During this period, a new officer came in to head Special Projects, and as well as I got along with his predecessor, I did not mesh well with him. Maybe it was because he had little use for golf but probably it was more that I started acting cocky and tended to ignore some of his instructions. I did however have a great relationship with the Special Projects Assistant Department Head Tom Seymour, who in is spare time wrote poetry and dedicated one of his efforts to me. A short poem of his is shown below,

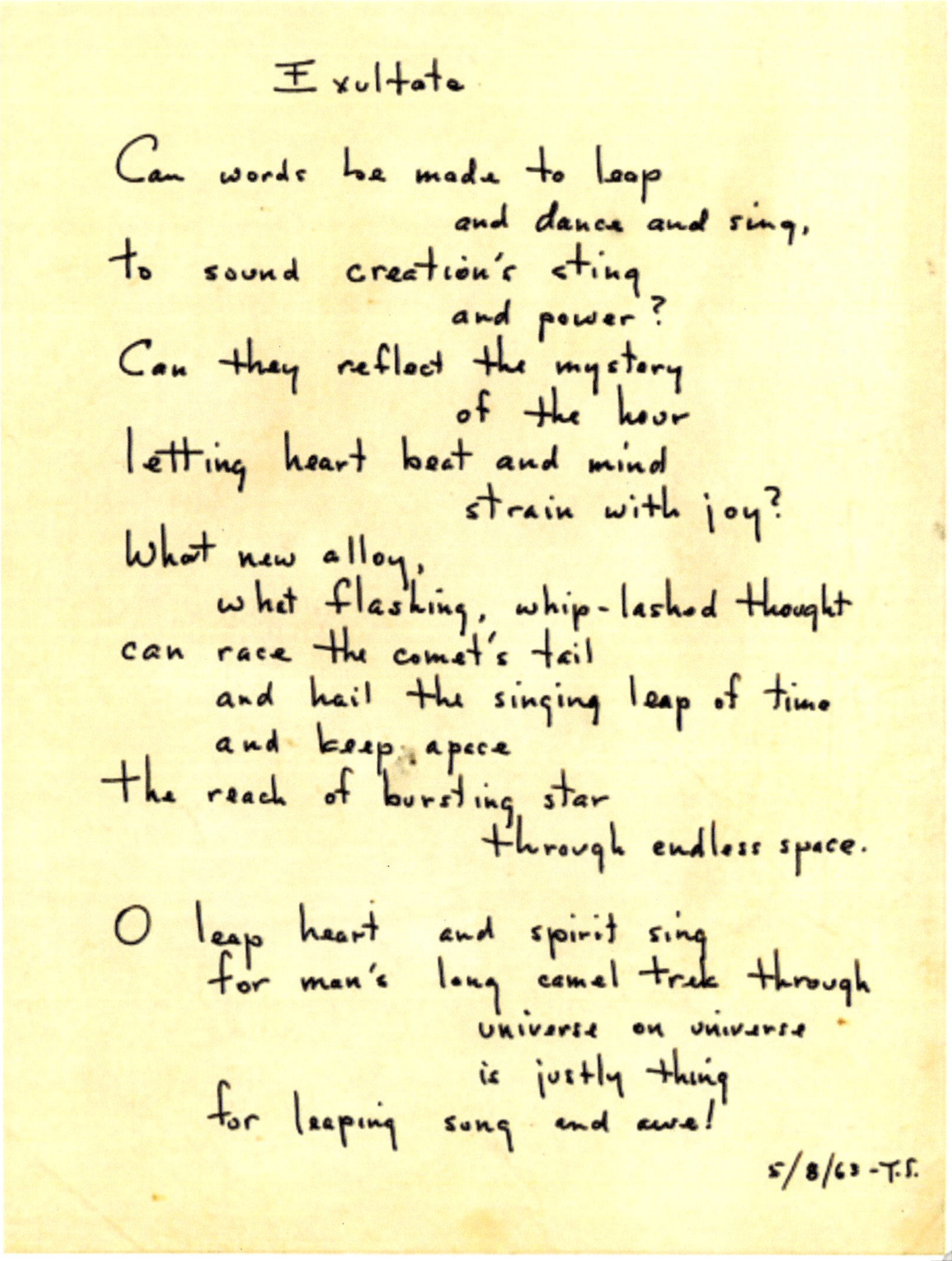
Exultate

Can words be made to leap
and dance and sing,
to sound creation's sting
and power?
Can they reflect the mystery
of the hour
letting heart beat and mind
strain with joy?
What new alloy,
what flashing, whip-lashed thought
can race the comet's tail
and hail the singing leap of time
and keep apace
the reach of bursting star
through endless space.

O leap heart and spirit sing
for man's long camel trek through
universe on universe
is justly thing
for leaping song and awe!

5/8/63 - T.S.

Cuban Missile Crisis

In the fall of 1962, Fidel Castro, Nikita Khrushchev and medium range ballistic missiles took control of my life, and before long I was working 12 hour days (14 hours if one counted turnover time), seven days a week. While Soviet Medium Range Ballistic Missiles (MRBM) arrived in Cuba in mid September, it was not until October 14th that a U-2 reconnaissance mission discovered a site at San Christobal. This precipitated in short order: a naval blockade (quarantine), invasion force build up, a U-2 shoot-down and a rash of reconnaissance flights over the island, both high and low altitude.

To give a sense of how dire the circumstances at the time were, the JCS instructed the Strategic Air Command to go to Defense Condition 2, the only confirmed time in history. The NAVPIC role in the imagery reconnaissance exploitation effort was secondary to the National Photo Interpretation Center and complementary to the Air Force effort run out of Shaw AFB in SC. While the first evidence of the missile sites was obtained from high altitude U-2 imagery, the low level tactical recce photos were vital in convincing doubters when the question was asked: "are you sure?."

On the right is a photo of a MRBM missile site taken by the forward firing camera of a Navy Crusader F8U-P. To get the imagery into Washington DC, one of my missions was to arm myself with a 38 S&W revolver, shoulder holstered over my dress khakis, drive out to Andrews AFB to meet a T-34 courier aircraft from Homestead AFB FL, and take custody of a crate of film. Then I would drive it to the building housing the Steuart Motor Company in NW Washington DC to be interpreted. Today it is hard to believe that the top four floors of this building, then located pretty much in the middle of an urban ghetto, housed the best photo interpreters in the world. Charged with exploiting this imagery as well as photography from other sources, much of whose existence could not even be revealed, the national center did a fantastic job in correctly and repeatedly identifying the highest threats to our national security.

There is a good short history of the Cuban Missile Crisis on the web at https://www.cia.gov/library/center-for-the-study-of-intelligence/csi-publications/csi-studies/studies/unclass1992.pdf. Put out by CIA as part of its *Studies in Intelligence* series, see the section on The San Christobal Trapezoid by John Hughes. I note this particular article because he would later become my mentor and perennial employer. Thanks to DIA, you can see the briefing John gave about the crisis on television at http://www.dia.mil/history/features/cuban-missile-crisis/

Post crisis I got to brief one of President Kennedy's entourage, Under Secretary of the Navy, Paul Fay. I spent a lot of time preparing the presentation to make sure I used every minute productively. I didn't get the first sentence out of my mouth before he interrupted with a question, and the briefing never got on track after that. He seemed to believe he knew more than anyone else in the room and didn't need assistance in understanding what went on in a very complicated environment. The result left a bad taste in my mouth and disdain for the whole class of political hacks appointed to responsible jobs in the Defense Dept.

One benefit of all of this was the Center was awarded a Navy Unit Commendation (shown below) for its work and I finally got to wear a ribbon. A copy of the citation itself along with others received over the years is contained in Appendix A. Around this time, I also was authorized to wear the National Defense Medal ribbon.

United States Navy

Lieutenant Francis A. BAUSCH, USN

is hereby awarded this

Certificate of Commendation

based upon a

Navy Unit Commendation

for meritorious service with the

Naval Photographic Interpretation Center

Washington, D.C.
28 May 1963

Captain, U.S. Navy
Commanding

The New Year rolled in peacefully and I got back to my original charter to produce PI reports using ADP techniques. I had mixed success but I did get to understand the in and outs of Electronic Accounting Machines, then the state of the art in Information Technology. I also got involved in secure communications as I was the action officer for sending extremely sensitive information over to the Pentagon via telephone. In one of the basement vaults we had a paper tape, cipher generator which randomly encrypted the text for transmission. I would sync up with my counterpart in the puzzle palace and off the information would go. This method is essentially fool proof as each message gets a completely new, random cipher key when it is sent.

And should one think that it was all work and no play, my wife and I did get out to socialize every once in a while; a photograph of o 1962 Christmas party thrown for the Center personnel is shown below. In the photo Mary and I are talking to CDR Smith, the Executive Officer of the Center, and his wife. While we were basically stay at home people, we did very much enjoy socializing with my co-workers and we became close friends with my boss Jim Mickelsen and his wife Donna.

Early in 1963, I got a great break concerning my next assignment. Involvement in the Cuban missile affair had put me in routine contact with Lt. Bob Morsches on the CINCLANT/CINCLANTFLT staff in Norfolk VA and in June 1963, based on his recommendation, I took his place as the Special Activities Officer in the Intelligence Division. Given the fact that Norfolk was just a couple of hundred miles down US 17, you would have thought the move for the family would have been a breeze. But it would take temporary stays in Poplar Halls and a beach rental on Willoughby Spit as we played vagabonds over a number of months before getting settled. But when we did, we really hit the jackpot in the way of house and location. This makes for an interesting story of its own, my buying the house on Sea Chest Lane in Baylake Pines sans consultation with anyone including my wife. Needless to say, I never bought another house without checking first with Mary.

After a going away lunch that obligated me to drink a martini for each of the attendees (I think I got to six), all the golfers went out to play nine holes of golf at Andrews AFB. I remember falling off the elevated 10th tee after a topped drive to the hilarious laughter of my fellow spooks. Shown below is the going away poster from the Special Projects officers that brings back many great memories.

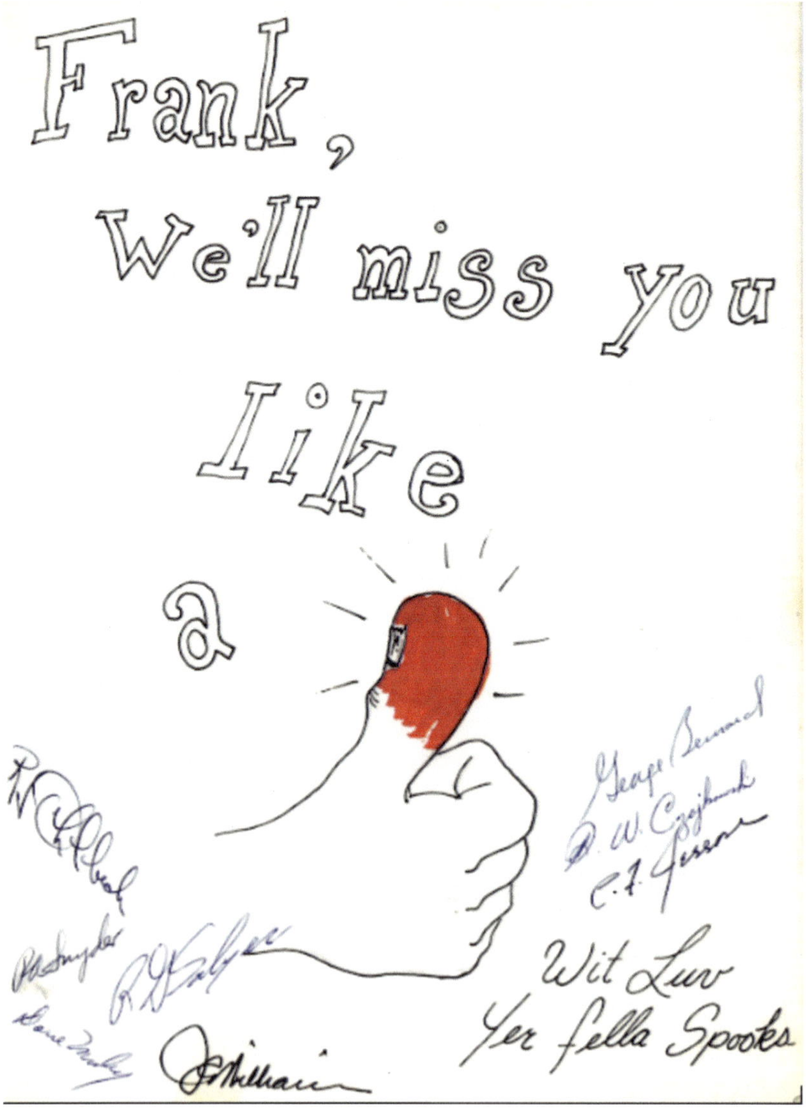

Chapter 4–Atlantic Fleet (1963-1967)

Staff, CINCLANT/CINCLANTFLT

The combined headquarters in Norfolk VA of the Commander in Chief Atlantic Command (CINCLANT) and the Commander in Chief Atlantic Fleet (CINCLANTFLT) was situated in a compound halfway between the Naval Station and the Sewell's Point Golf Course. This compound also headquartered the Supreme Allied Commander Atlantic (SACLANT) which had some recreational relevance to me. Its head, Admiral H.P. Smith, was the same big boss I had at CINCLANT but he had an entirely separate staff which I never had to interact with professionally but it had an officer's mess where long lunches washed down with cheap drinks inevitably led to an afternoon of golf. While this combination often turned into a career ender for some officers, I would partake sparingly of the lunchtime spirits but I always was available to play golf. I basically enjoyed the interaction with this expanding circle of NATO colleagues more than the free time and I mention this at the start of this section to let the reader know that the level of stress for this new job was generally minimal. As you will see, only a few times did it get interrupted by intense crisis situations.

This tour provided my first contact with the 1630 designation naval intelligence community, notably J-2 Capt. Grumpy Blonts and his successor, Capt John Q. Edwards. Both were real jewels and instrumental is getting me ahead. I was the Photographic Intelligence /Special Activities Officer for the staff and had a Chief Petty Officer assistant to handle the paper work. Basically my own boss in the Special Activities area, I built slick, desk side type presentations using the latest satellite imagery (high resolution from space became real in 1964) and regularly briefed the CINC, Vice CINC, J-3 Ops and J-5 Plans, all of whom were cleared. I am not too sure how much they absorbed but they were pleased that someone in J-2 came around and shared these goodies with them.

I was also given the responsibility for reviewing submarine reconnaissance reports and I interfaced with the nuclear targeteers, semi-sharing a secure space with them. As a result I got to meet a lot of the Strategic Air Command (SAC) and key Joint Strategic Target Planning Staff (JSTPS) people. On one winter flight to Omaha in a Navy Beechcraft, watching ice build up on the wings made the trip exciting as well as unforgettable. Equally memorable for different reasons was another Beechcraft fight, this time a summer one with pilot friend LCDR Bob Heon to NAS Quonset Point just south of Boston on reserve intelligence training business. Essentially a cover to visit his cottage on Cape Cod where his family was staying, I marveled at the sheer audacity of this adventure but equally so at the weather on the Cape that was windy and cold, resembling Alaska more than any beach I had visited up to that time.

My immediate boss was Army MAJ Leon MacMullen who was keen on using all fleet assets for intelligence collection. He was a veteran of the WWII landings in Italy, a great teacher and had absolutely no reservations about going into a navy environment and telling sailors what they had to do in respect to shipboard collection. He and I spent a lot of time with PHIBLANT, MINLANT and CRUDESLANT staffs selling them on using their deployed units to send in Information Reports as a way of life.

I was promoted to LCDR in July 1964; the photo below shows our family of seven at the ceremony. It looks like only five but John is hiding between Jim and Ann on the right (you can see his legs if not much else) and Frank Jr. is tucked away in his mother's womb. His birth was one example of a hectic time I had in Norfolk. Standing Command Duty Officer watch at the operations center, an early morning call from home that 'it was time' got misrouted, and we got a late start to the hospital. After loading everyone in the car, I got a flashing red light police escort through Norfolk to the Portsmouth tunnel, a short distance from the Naval Hospital, where we arrived just in time. But there also was also one equally exciting but more work related situation.

Dominican Republic

In order to avoid what he thought might become another Caribbean crisis like Cuba, and acting on the advice of Ambassador W. Tapley Bennett, in late April 1965 President Johnson ordered US forces to intervene in the Dominican Republic to 'calm the situation.' The government there had fallen apart and rebel groups were threatening life and property. My part in all this stemmed from the fact that CINCLANTFLT was directed to deploy an amphibious squadron off the Dominican coast and throughout the entire undertaking, CINCLANT exercised operational control over all US Forces in the area. For roughly a month, I stood watch in the operations center mostly monitoring developments and trying to make sense of how the Command and Control apparatus functioned (or in this case malfunctioned). In the space of the first week, US nationals were evacuated, the marines were landed, and the 82nd Airborne Corps flown into an airfield near the capitol Santo Domingo and soon thereafter unleashed on the countryside. The Organization of American States got involved and within a month's time, an Inter-American Peace Force headed by a Brazilian general was formed to take over the US mission. To give you an idea of the scope of the effort, in mid-May almost 24,000 US soldiers, marines, and airmen were in the Dominican Republic, some 40 odd naval ships were positioned offshore, with other Navy and Air Force units were deployed close at hand in the Caribbean. These forces assisted in the evacuation of nearly 6,500 men, women, and children of 46 nations, and distributed more than eight million tons of food at a cost of 24 American servicemen killed and another 156 wounded. One of my more unnerving tasks was trying to patch a phone call from President Johnson direct to LGEN Bruce Palmer, the onsite Commander (COMDOMREP). While the chain of command theoretically went from the President through the SecDef (McNamara at that time) then through CINCLANT down to General Palmer, President Johnson was too impatient to let it work that way. And you can imagine all the complications that kind of shortcutting caused.

Professional Growth

While I was generally kept busy on the staff, there were no challenges, and thus I made no big mistakes. My family prospered. The kids very much enjoyed the beach just a few blocks away from our house; on the right is the original photo on a 1964 Christmas card photo we sent out which shows a lot of happy faces. The fact that the tour was credited as "sea duty" was a bonus and in way of further bonus, two carpoolers opened my eyes to a hitherto unknown world of combat. CIA Liaison Officer (CAPT Jim White) and the Psychological Warfare Officer (Col Ed McLean) both lived near me in Baylake Pines and most days we drove back and forth to work together. I was a better listener than a

talker then (I wish I could have stayed that way) and I absorbed a lot about the unconventional aspects of war in the process of our travels. In addition to saving gas as well as wear and tear on my car, carpooling sometimes left me on my own to get back home from the Little Creek golf course following an afternoon round of golf. And what a magical thing it was to be picked up by wife and kids at the golf course in our Mercury Colony Park station wagon. This high spirited mission of course was to bring Daddy home but more often than not, it might turn out to be a late afternoon trip to the beach, or better still, to a drive-in-movie which we would all sleep through, or some other great adventure we would dream up.

And while I was totally enjoying myself, my boss continued to take an interest in my advancement. CAPT Edwards encouraged me to apply for a change of designator to 1630 (Intelligence), basically saying there was no future in remaining a 1350 (Aviation Ground). I did as suggested and while that particular request was denied, I reapplied the next cycle recognizing that I just might have a career in the navy. CAPT Edwards also put me in for a Joint Services Commendation Medal (JCSM). I was detached from the staff in late September 1965 after a little over two years, to first attend a technical training school and then report for duty aboard the aircraft carrier USS Independence where the JCSM would be presented in a ceremony on the hangar deck.

NAS Sanford (IOIC School)

My orders to the Independence designated me as the prospective Integrated Operational Intelligence Center (IOIC) Supervisor with a reporting date of December 1965. I spent eight great weeks at NAS Sanford FL learning the ins and outs of the Integrated Operational Intelligence System (IOIS)—the combination of airborne collectors and shipborne processors. The Naval Station, now the municipal airport for Sanford, was located just north of Orlando and I received personalized instruction being a one-man class, attending many lectures with my teachers out on the Mayfair CC golf course. But I was a conscientious student and when not in class, became fully involved in examining one of the major difficulties thwarting the efficient functioning of the IOIS, something called the code matrix block. Working with instructor LT Bill Damon, instead of imprinting a code matrix block image containing relevant navigation and sensor data on the film, we proposed using a central data recorder (CDR) in the aircraft and simply time tick the film for reference back to the CDR which would contain this data in digital form. In this way all sensors could also be correlated to each other using the CDR plus if a correction was needed to the data, it could easily be applied to all sensors. The major problem with this approach, other than a large retrofit, was that the Air Force did it that way. And admitting the Navy might have been wrong and the Air Force right was not something the IOIS sponsors in Washington wanted to do. In any case, the report was professionally done and while we felt bad about its poor reception, we recognized we did what we were supposed to, i.e., try to make the reconnaissance system work better. But unfortunately, the small amount of notoriety that accompanied our work singled me out as an IOIS critic, something that would be reinforced during my tour on the Independence.

USS Independence (CVA-62)

I finished up at Sanford in time to spend Christmas at home, and reported to the Independence to relieve CDR Frank Hofer in January 1967. The ship had just returned from a Western Pacific cruise where it had operated off Vietnam. Our AirGroup (CVW-7) as well as the reconnaissance squadron RVAH-1 were the same ones the Independence (shown in the photo below) had on board during the cruise. The IOIS had been fully tested and found wanting in many respects. The RA-5C Vigilante reconnaissance aircraft was a tricky plane to fly (two were lost on the WestPac cruise) as well as a maintenance monster, while the shipboard equipment, especially the Stereo Comparison Viewer did not work very well. When push came to shove, imagery taken over Vietnam was usually viewed using an ordinary light table. While this and other equipment I inherited was suspect, the officers and 50 enlisted working for me were top-notch people. (photo on the right shows some of the ship's intelligence officers assembled on the flight deck—left to right, top row, Woody Wood, Dick Zolman, Jay Stevens, myself and CAG AI Elton Johnson; bottom row, kneeling, John Sanger and Brendan Bart.

The ship did the usual work-ups with the Air Group while I continued to broaden my expertise by attending the Electronic Warfare training course offered at Dam Neck VA. In June 1967, we deployed to the Mediterranean for duty with the Sixth Fleet. One of my first responsibilities when we got to the Med was briefing visitors about how the ship operated off Vietnam. I noted of course that I had not been there but explaining how things worked was easy and after a while I didn't even caveat that many of the events I was recounting occurred when I was playing golf at the Mayfair CC.

Bridge busting has always been a highlight of naval strike aviation (read *The Bridges at Toko-Ri*) and I was called on to relate how our Air Group executed its most notable attack in Vietnam on the Thai Nguyen Bridge. The main body of the October 1965 strike on this target located north of Hanoi deep inside North Vietnam was composed of four A-6Es each with five 2000 lb bombs; four F-4Bs with rockets or four-1000 lb bombs; 16 A-4Es with one-1000 /two-500 lb bombs; and four F-4Bs with AAM/Zuni rockets or one-1000 lb bombs. Despite the large amount of ordnance dropped on it, the bridge remained standing after the raid while the ship suffered the loss of three Phantoms, with two aviators killed and four POW'ed. Another notable Vietnam carrier strike on a bridge, conducted by a sister ship, was captured in the colorful painting "Dragon Slayers" by Robert Bailey pictured below.

This attack on the Thanh Hoa Bridge is of special interest to me as the artist identifies the lead F-4 pilot flying the flak suppression mission as Fred Ferrazzano, a kindred spirit from Paterson NJ and shipmate of mine on the 1956 Lexington Cruise.

As for operating the IOIC, even with ten years of navy experience and confident that I knew everything to know about the system, nothing could prepare me for what happened during the cruise. To compound the problem, right off the bat, I got into a shouting match with the CarDiv Intelligence Officer during the en route ORI (Operational Readiness Inspection), something that soured our relationship for the entire cruise. But that turned out to be the least of my problems. Before I get into that, first a little bit about what liberty was like visiting ports in the Med.

We relieved the USS America, anchoring with her off Palma, Majorca right before she headed home. That morning's weather provided a unique visceral Mediterranean experience: low clouds, a misty haze as warm, crisp, fresh air engulfed the entire bay with an eerie stillness, a *South Wind* type atmosphere. Over the next seven months, we visited Majorca, Cannes, Barcelona, Naples, Argostoli Bay, Istanbul, Malta, Beirut, Genoa, Palma and Livorno where I'd sightsee as much as possible (a trip to Rome being the highlight). During each of the port visits, I would write a post card to one of the kids with some words of wisdom; the postcard photo on the right shows the Galleria in Naples with the message to John "hope you are being a good boy for Mommy and do as she says." I was a part of our skipper's (Blackie Kennedy) golfing foursome and for the first part of the cruise, more often than not I was 'forced' to play almost every in-port day. And what a strange golf environment it was: barefoot teenage girls caddying at the Cannes-Nice-Antibes Golf & CC; wizened women in black descending on the green between foursomes to remove weeds by hand at El Prat west of Barcelona; British oil tanker captains hanging out at the Zahrani G. C. clubhouse in the hills over-looking the Tapline terminal just south of Sidon, Lebanon; and a round at the venerable Istanbul Golf Club (4th oldest in Europe) where I fit in a ferry trip across the Sea of Marmara and a bus ride to visit the Naval Security Group's facility at Karamursel.

But nothing hit the spot like a letter from home. I have kept all my wife's letters from that period and other than rating me a five star husband, I will not share their contents with you. But they show how a remarkable woman caring for five young children, handled seven months on her own like Superwoman.

IOIC Problems

The problem with the IOIC was the kind that I could never get ahead of and it got worse in phases. Phase 1 began with both aircraft sensor and ship processor equipment failures of an unprecedented nature and rate (at least I thought they were unprecedented) and having to use a data base that had numerous errors and was horribly packaged. Phase 2 was characterized by numerous VIP tours where we tried to show off 'our new way of doing things.' But when we fired off the equipment, power failures or other glitches would occur just as we began to demo the wonders, for example, of the Stereo Comparison Viewer. With 100 hours of maintenance required to keep one of these monsters operational for five minutes, crew morale sank; 16 work hours each day never seemed to be enough. And then the final phase of the disaster; in desperation, after getting an okay from the skipper and ops officer, I ratted out the system to the Navy IG. He inspected Independence midway into the deployment, and as the former Vice CINCLANT he remembered me, and when he got back to D.C, he did what I wanted him to—raise hell. I have no idea how much it helped if any, in getting the program squared away, but feedback from Tony Sesow in the Program Officer indicated it almost got me yanked off the ship.

Probably the fact that the cruise was almost over when this came to pass and I could do little more damage, saved me from that ignominy. Looking at my fitness reports for this period, all of which were two-blocked left, one would never know any of the above. Very supportive bosses and coworkers such as CAPT John Fox, who took over the ship midway during the cruise, Operations Officer CDR Swoose Snead (later a RADM, CDR (later Capt) Barney Martin on the Sixth Fleet Staff, Assistant IOIC Officer LT (later Capt) Dick Zolman, and a crew of superb DS and MAs, were all key in getting me get me through this very troubling time.

Like many of my other tours of duty, my immediate superior was key in my career development. The very direct and useful guidance the "Swoose" provided me about writing fitness reports was very helpful. I had done drafts for his signature that he was not satisfied with and he called me in one day and told me to decide for myself whether or not the subject should be promoted. If 'not' he said, write an objective report. If 'yes', make every effort to impress the reader and try to make the officer you were reporting on sound like 'god incarnate.'

I noted above that we had a lot of VIPs tour the intelligence spaces (Ambassador Chip Bohlen for example wanted his French guests to see the great satellite photography we were collecting). But there were a lot of more serious interactions with some of our NATO allies that indirectly related to this imagery. The photo on the next page was taken during a nuclear strike briefing to the British Minister of Defense (MoD).

A major part of our cruise was spent preparing to execute the NATO Strike Plan where the command and control procedures were complex. When the Independence moved into the Mediterranean, it essentially became a NATO nuclear strike asset under the operational command of the Supreme Allied Commander, Europe (SACEUR). In respect to the control of nuclear weapons however, it remained a part of the US command and control structure with its attendant safeguards against accidental or malicious release. Along with most of the LCDRs onboard, I was designated as an Emergency Action Message (EAM) responder, two of which would always have to be on board ship to react. I was made responsible for the coordination and scheduling which required a lot of work since 'test' messages would routinely be sent from the National Command Authority (NCA) to insure the system was indeed working. In a sense I became the EAM boss, specifying the procedures we would follow to authenticate the coded message plus insure the necessary internal ship procedures worked, like getting word to the proper people.

As for the strike operation itself, the US had years before developed a Single Integrated Operations Plan (SIOP) for nuclear attacks and much the same thing was accomplished with NATO weapons carriers. So the timing of the effort was not a problem. As for enhancing the chances of a successful mission, target folders were developed using US Secret material which was selectively disseminated to our allies. When satellite imagery was available for the targets the Independence was to strike, it was provided separately where I would keep it under control until we got the 'go code;' then I would make it available for pilots to use. To my mind this was an unsatisfactory arrangement as I could not conceive of anyone wanting to be bothered at the last moment attempting to digest new targeting materials as he tried to figure out the best way to drop an atomic bomb on an important target and get away safe. But that was only speculation on my part, as thankfully I never got the chance to authorize opening these 'sealed packages.'

Six months into the cruise, with the tempo of ops slowing during the Christmas holiday season, I opted to fly home on a chartered Air France 747 for Christmas. It was an expensive and short trip but came at a time my morale really needed a boost. A month and half later, the ship itself would return to Norfolk. Although I did not know it at the time, this would be my last cruise aboard a carrier and after twelve years of watching flight ops off and on, I thought them a special treat and still remember them that way. As we departed, there was some recognition for the work we had done; below is a copy of the message the ship received after we left the Med from Vice Admiral Ashworth, the Commander Sixth Fleet that made a point of highlighting our '*unusually alert intelligence effort.*' You can see my notation at the bottom right of the message reminding me who to thank for the nice words.

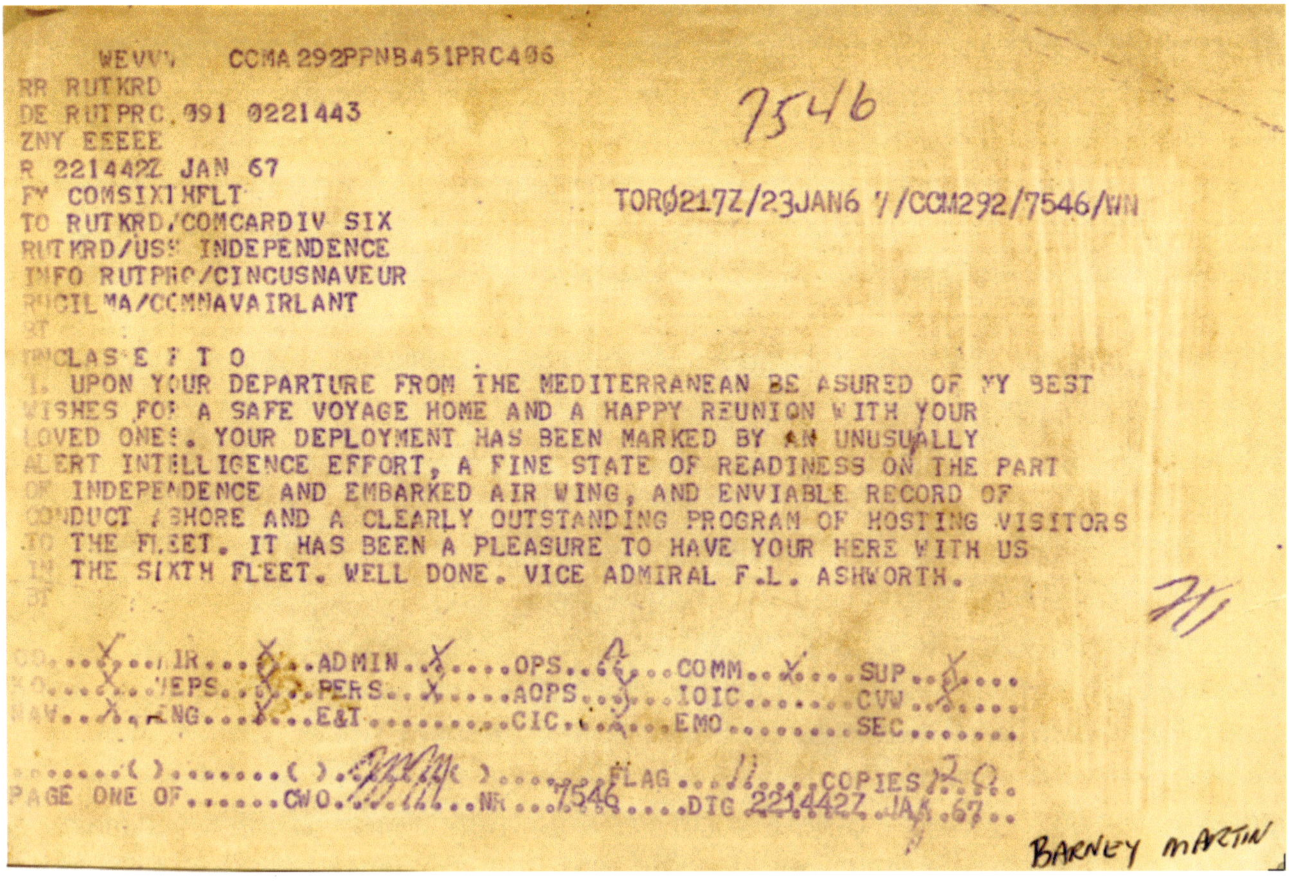

WEVVV CCMA292PPNB451PRC406
RR RUTKRD
DE RUTPRC 091 0221443
ZNY EEEEE
R 221442Z JAN 67
FM COMSIXTHFLT
TO RUTKRD/COMCARDIV SIX
RUTKRD/USS INDEPENDENCE
INFO RUTPRC/CINCUSNAVEUR
RUCILMA/COMNAVAIRLANT
BT
UNCLAS E F T O
1. UPON YOUR DEPARTURE FROM THE MEDITERRANEAN BE ASURED OF MY BEST WISHES FOR A SAFE VOYAGE HOME AND A HAPPY REUNION WITH YOUR LOVED ONES. YOUR DEPLOYMENT HAS BEEN MARKED BY AN UNUSUALLY ALERT INTELLIGENCE EFFORT, A FINE STATE OF READINESS ON THE PART OF INDEPENDENCE AND EMBARKED AIR WING, AND ENVIABLE RECORD OF CONDUCT ASHORE AND A CLEARLY OUTSTANDING PROGRAM OF HOSTING VISITORS TO THE FLEET. IT HAS BEEN A PLEASURE TO HAVE YOUR HERE WITH US IN THE SIXTH FLEET. WELL DONE. VICE ADMIRAL F.L. ASHWORTH.
BT

7546

TOR0217Z/23JAN67/CCM292/7546/WN

CO...IR...ADMIN...OPS...COMM...SUP...
XO...WEPS...PERS...AOPS...IOIC...CVW...
NAV...ENG...E&T...CIC...EMO...SEC...
PAGE ONE OF......CWO......NR...7546....DTG 221442Z JAN 67..

BARNEY MARTIN

End of Cruise

The Independence got back to Norfolk in February 1967; on the right the ship's photographer caught Mary and I leaving Pier 16 the day we tied up. Does it look like we are happy about being back together? Indy still had a few short operating periods off the coast but then went into the Portsmouth Naval Shipyard for an extended overhaul in March 1967. What a terrible time to be aboard a ship, when after working so hard to get everything to work right, you watch it all torn apart. I took advantage of the slack period and went to Nuclear Weapons Employment School where I learned how to optimize a given yield weapon against various targets.

That same month I was designated a Command Duty Officer (In Port). As insignificant as that is to being qualified as one underway, it is still is a lot of responsibility. There is no more hazardous place than a shipyard with combustibles and welders everywhere. On those nights when I had the duty, Mary and the kids would come for dinner in the wardroom. A proud moment to show off my handsome family with the only downside, Frank Jr. not liking to go down the hatches. And then mercifully, transfer orders came in August and we all moved to a rental house in Maryland where I would commute to my next duty station in Anacostia as a student in the Defense Intelligence School. My going away party was a memorable one, primarily because of a stupid move I made lifting a keg of beer out of a garbage can full of ice, something my back took note of forever. The Independence was decommissioned 30 years later and I last saw her in 2009 tied up to a pier at Puget Sound Naval Shipyard in Bremerton WA, where she awaits scrapping.

Chapter 5–Defense Intelligence School (1967-1968)

Post Graduate Selectee

In late 1966 while still deployed in the Mediterranean, I received news that I had been selected to attend Post Graduate (PG) School. This was a surprise but the 64 million dollar question at that time was where? I had dreams of MIT (how I could ever envision that given my past scholastic record is a mystery to me). As it turned out, it was the Defense Intelligence School (DIS) in Washington DC where I would attend the nine-month Defense Intelligence Course. The school was located on the Naval Station/Anacostia Annex, a stone's throw away from where I had gone to Photo Interpretation School back in 1955.

DIS has since changed both its name and location. In the 1980s it became the Defense Intelligence College; in the 1990s, the Joint Military Intelligence College; and just recently, the National Intelligence University; and it physically moved a short distance south onto the Joint Base Anacostia-Bolling as part of the DIA intelligence complex. I mention all this recent history because the University honored my perennial 'boss-to-be' by naming its library after him. The John T. Hughes Library houses over 2.5 million items to include books, unclassified intelligence documents, reference materials, periodicals, microfilms, video, maps, charts, international periodicals, newspapers, annuals, serials, and statistical reports. Russian periodicals. microfiche and microfilm collections of general and scholarly periodicals, FBIS reports, and declassified documents can also be found there.

At about the same time I learned about my selection to PG school, I was selected for change of designator to 1630, which I accepted. Looking back at it now, it does seem somewhat ironic that just having been certified an intelligence officer, I was now going to school to find out for the first time what the general field of intelligence was all about. But the real miracle was that I got selected to go to DIS at all. I had a long history of mediocrity in school and my test scores across the board were pedestrian. I was a C student in college, my Officer Qualification Test had me at the 50 percentile, my OCS ranking was in the bottom third and Aviation Ground Officers School grades put me right in the middle. I did a little better in schools where I had some experience but still had problems. For example at Alameda in the Air Intelligence Course even though I was an instructor in the school, I barely ranked in the top 10%; and in the Non-Resident Defense Intelligence correspondence course, I just made the top 25%. Imagine having a hard time with open book tests. While I did get some of A's during the early 1960s in the Military Studies courses at Maryland, I was never at or near the top of my class. In general, I never thought of myself as anything other than an academic misfit. But all that was about to change.

Defense Intelligence Course

I would not surprise the DIS school officials as far as grades were concerned however; after nine months of instruction and probably the most experienced officer in the course, I just barely squeezed into the top 10%. I could not blame my poor performance on anything other than lack of application. There was no pressure to do better; life at home after the move from Norfolk went very smoothly with us finding a great place to rent in Belair (Bowie MD) where my family thrived. I carpooled with a neighbor, MAJ Frank Blackburn, which eased the travel burden and we productively used our trips back and forth to go over what we had picked up in school. And although my class standing never improved, inexorably I started to change my way of thinking about learning. While I continued to receive average grades, I got to the point where I recognized finding out about things was exciting and in many cases beneficial. And this learning curve was steep as only four years later I would graduate from the University of Maryland with a Masters in Government and Politics earning top grades and the respect of my grad school peers and professors. How that all transpired was very much the result of the nine months I spent at DIS.

While I may have been blossoming as a student, it became harder and harder for me to toe the line in respect to the school administration and their expectations for me as a military officer. Being a LCDR I was put in charge of a section of 15 other student officers and asked to supervise them. They were a great group of guys who needed no supervision and even though they mocked my past PI experience with a special presentation when we graduated (see cartoon on right), I loved them all. But being Section Leader from the school standpoint meant that when something went wrong, I would have to answer for it. I never told the school officials this was something I thought inappropriate and instead let my actions speak for themselves (mostly ignoring guidance from on high) and while I probably could have handled it a little more smoothly, they grudgingly accepted me as a student rather than as a glorified hall monitor.

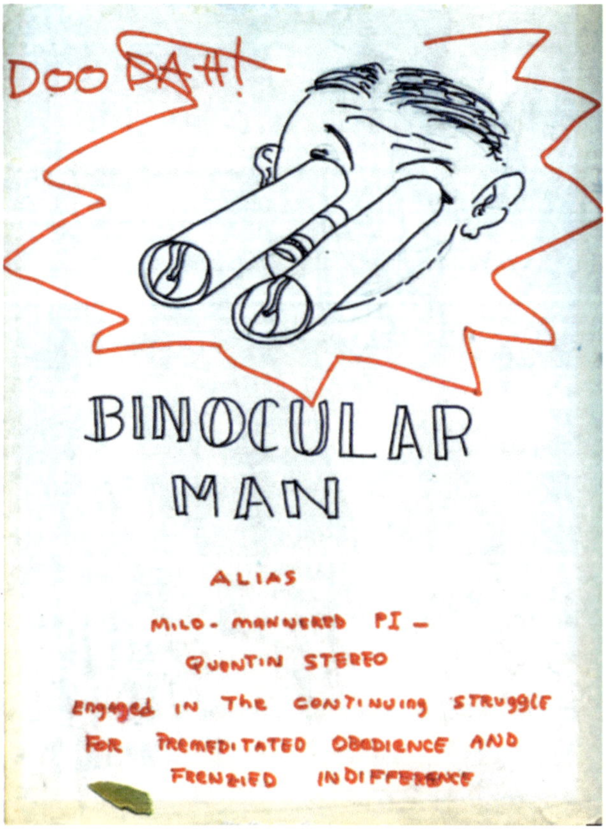

Instructional Staff

The DIS staff was well rounded and had some real gems; Abe Kalish was a great guide to researching and Frank Romance, an eloquent speaker on International Relations. But the real gems of the school were the invited guest speakers. Even the mundane subject of economics was treated with gusto by a US Military Academy traveling troupe that put real meaning and even drama into the circular flow model. And so it went with most all the speakers. The one that sticks out most in my mind was Karl Wittfogel, author of *Oriental Despotism*. A white haired, shambling giant, his speech was barely intelligible and his book turned out to be no easier to decipher. But just listening to him and his passion for the subject, you knew that the matter was vital and worth whatever it took to better understand. I've had the book out from the library a number of times and still have a hard time with it. Nonetheless his articulation of the concept of 'hydraulic society' and the way he formulated it, is one that will never leave your consciousness when you discuss governmental forms of any kind. There were many other great speakers, all whom seemed to open doors to learning in a marvelous and mysterious way. Then 34 years old, I started to significantly expand my personal library, something that helped further open the doorway to knowledge and understanding. And it became very clear to me that if I were to get anywhere in the future, having an advanced degree would help immeasurably. Within months, I would become a graduate student at the University of Maryland looking to get a degree in the general field of International Relations.

Back to the Future

It would be some 12 years later, after I retired from the Navy, that DIS would again become part of my life. During my last active duty tour on the Intelligence Community (IC) Staff, one of my shipmates was MGEN Jack Thomas, USAF (Ret.) a former Assistant Chief of Staff of Intelligence. We both left the IC staff at the end of 1978; I went to work for John Hughes as his Special Assistant and Jack became a full-time consultant in the Office of the Secretary of Defense. Part of his portfolio was to conduct a graduate seminar at the Defense Intelligence School entitled "The Development and Future of the Intelligence Community,' and he chose me as one of his guest speakers. I would regularly venture over to the same building I took the Defense Intelligence Course and talk about a subject I then knew a lot about—the National /Tactical Intelligence Interface. Jack was well respected as an intelligence professional, and so much enjoyed the work, he hung on consulting for OSD until he was in his 90s (he died in 2008 at age 98).

Below is an excerpt from a letter he wrote to Mr. Hughes re my efforts, which may give you some sense of my transformation from a poor student to a quasi-professor.

I don't know anyone who was been more involved in the national/tactical intelligence interface... and this is why every time I conduct the seminar, I ask him to participate. He is highly effective both as a speaker and in the give-and-take environment of a seminar. His presentations stimulate questions, and my students were very interested in what he had to say, so they had numerous questions. Both Frank and the students...enjoyed the interchange.

As an adjunct on this matter, I have no reason to believe that he got anything out of my presentation but one of my more illustrious seminarians was future General Mike Hayden who eventually became (in sequence) the Director of NSA, Principal Deputy Director of National Intelligence and CIA Director.

Moving On

While my lack of enthusiasm for enforcing military discipline did not go over too well with the school administrators, I impressed them otherwise because at the end of the course, I was asked to stay on as an instructor. This was something I rejected out of hand and not too tactfully. I was looking forward to a tour where real things happened and sure enough, when I had my next set of orders, I would not be disappointed. After the workaholic, madcap experience onboard the Independence, the nine months at DIS was great in that it rested me up mentally for the all consuming job I was about to take on. Coming up was the first of my many associations with Mr. John T. Hughes.

Chapter 6-Defense Intelligence Agency (1968-1971)
Special Activities Office (DIAXX)

Coincident with orders to report to the Defense Intelligence Agency (DIA), I moved my family from Bowie to just outside the Beltway in Friendly MD where we bought a brand new house. Following closely after this move was the birth of our sixth child, daughter Susan Alexandra. I was assigned to the Special Activities Office located in the Pentagon and while not aware of the job specifics, just by the office title I knew I would be back working the national reconnaissance problem. The three years I spent in DIA may have been the most productive and formative of my life. This period assured a successful navy career (defined by promotion to Captain) and gave me the bona fides for another 13 years of post-Navy retirement work in government and industry as a 'reconnaissance expert.' It also marked my total separation from the real navy after 12 years of fairly direct involvement. And probably more important, it reinforced my work ethic as well as my understanding of what success meant, something that helped me adjust to some really difficult times to come.

DIAXX was an organization created around one man, John T. Hughes. And strangely enough, he didn't even head the office. As previously noted, he had become widely known in the intelligence community in 1962 when President Kennedy okayed him to give a presentation, broadcast nationally by television and radio networks, about the Cuban missile crisis. His obituary in the NY Times noted parenthetically that he had also personally briefed Presidents Ford, Carter and Reagan on Soviet military and other sensitive security matters. DIAXX had roughly 100 people and was headed by Air Force COL Frank Cappelletti, with a Navy CAPT as his Deputy and Mr. Hughes as his Special Assistant. The Office had four divisions with people assigned in various locations in the Washington DC area, all with a specific mission to fully control within the Defense Department the access to imagery products of the National Reconnaissance Program (NRP). And when I use the word control, I mean it in the broadest, most encompassing sense to include librarians, briefers, communicators, security officers, planners, interagency representatives, requirement managers and key exploiters. If you wanted to use an NRP imagery product in the Defense Department, you had to deal in one way or another with DIAXX. Why you may ask, and the reason was security. Until the early 1990s, even the fact that the NRP existed was classified Top Secret, special access required, on a 'must know basis.'

Organization wise, a Defense Department official, usually the Under Secretary of the Air Force, headed the National Reconnaissance Office (NRO). The NRO provided oversight for four reconnaissance program offices: A-Air Force; B-CIA; C-Navy; and D-Airborne assets. NRP products were exploited by specially cleared and trained analysts—Signals Intelligence by the National Security Agency (NSA) and imagery by the National Photographic Interpretation Center (NPIC)—but for the most part, the analysts had no access to collection system particulars. Of course one would have to be an idiot not to know the photos you were looking at, or the telemetry you were deciphering, came from space systems. But the operational parameters of the collection systems themselves were bound up in a separate security compartment. One of the few instances where analysts did have some insight into the working of the collection system, the KH-11 User Manual, was compromised early on when it was sold in 1976 to the Soviet Union by CIA analyst William Kampiles.

And like many high intensity, ultra secret operations with a critical mission, levity still had its place. On one of my trips to the NRO Staff Director's office (4C1000), I remember seeing friend Colonel Herb Tuttle's faux army recruiting poster on the wall of his office. Beautifully done in color, it showed a Polynesian girl in a sarong on a South Seas Island with the following text: "Join the Army: travel to exotic, distant lands; meet exciting, unusual people and kill them."

Reconnaissance Division

I checked into DIA on the last day of May 1968 and was assigned to DIAXX-1, the Reconnaissance Division. My boss, Air Force COL Ned Humphreys, was a complex character who was pretty abrasive and verbally abusive at work but a charmer outside of the office, something you could clearly see when he was in company of his wife Marcia. My wife and I attended a party at his house and I have never met a more gracious man, very much a Jekyll and Hyde kind of thing.

Our office, inside the National Military Command Center (NMCC) next to the Joint Reconnaissance Center (JRC), required three separate security check access points to enter. Multiple badges chained around your neck was the order of the day. The space was essentially a bowling alley with the boss in the pit (he had a door) and seven desks (secretary and six action officers separated by partitions running the alley's length. I was farmed out as an apprentice to Air Force LCOL Buzz Brudzinski to learn and then help in the satellite tasking business.

The basic document we used for requirements was the DoD Imagery Requirements Objectives List (IROL), which identified every single target in the world of interest to the DoD. Later I would take a cudgel to the IROL practitioners and try to get them to make it a more practicable document for satellite tasking rather than a broad requirements wish list. Even though the objectives were prioritized, the limitations of then current satellite systems required judgment that Buzz and I were supposed to provide on what specific targets a particular satellite mission would be directed against. What we would do is scan the IROL, extract specific objectives, put them into a list and present them to the United States Intelligence Board (USIB) Committee on Imagery Requirements and Exploitation (COMIREX) sub-committee that had responsibility for tasking the imagery satellite. This list would be broken down into two categories: search areas for low resolution, wide area coverage; and specific targets for the high resolution, postage stamp coverage. Then at meetings of the ICRS (Imagery Collection Requirements Subcommittee) who passed the requirements onto the NRO, we would sit down and horse trade, threaten, cajole, whatever was necessary to get our objectives approved for collection. A byproduct of this exercise was gaining a better understanding of how the collection system worked and eventually I wound up being a tutor for my bosses on things like orbits, ground coverage, etc., which was useful when issues related to launch time, inclinations, etc. cropped up. I used briefings slide like the one below to help them understand what was going on.

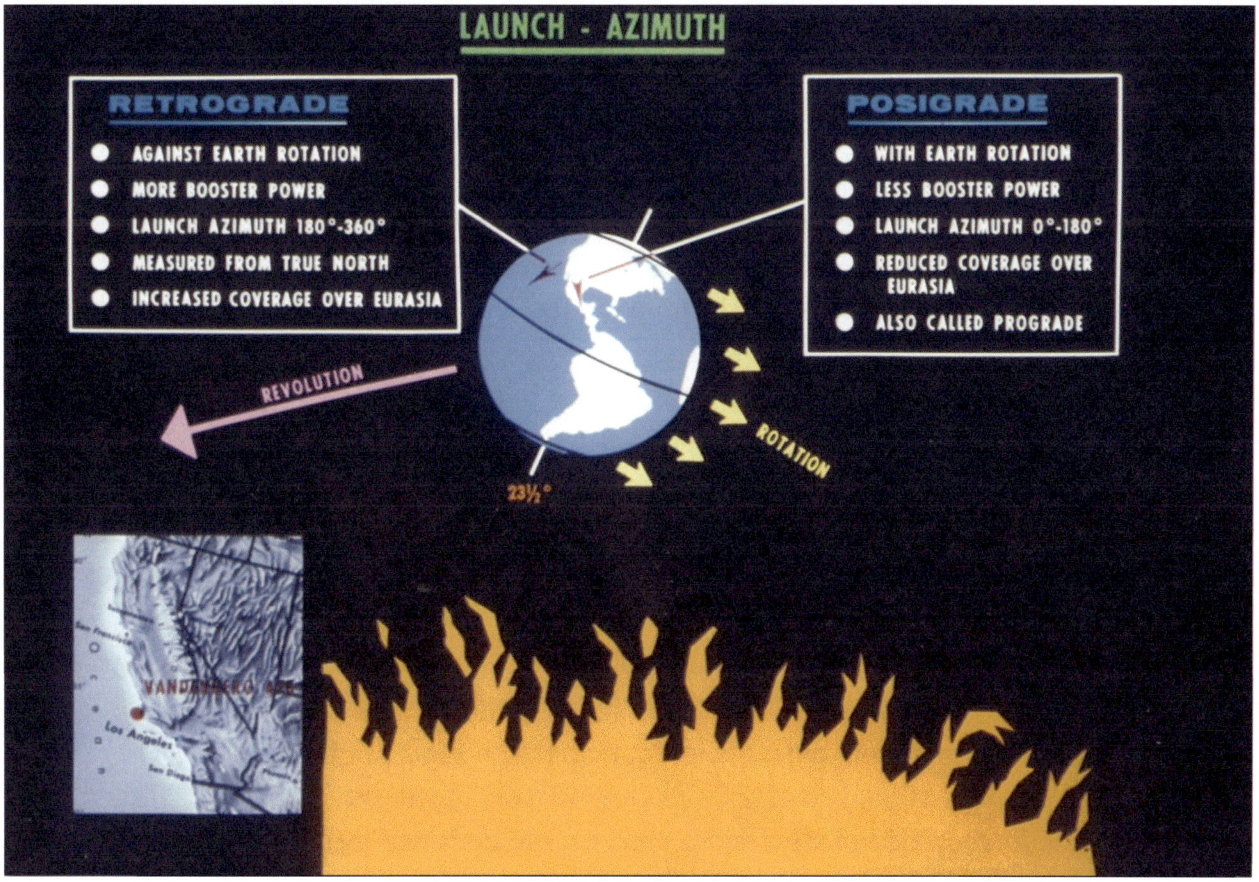

Moving up the Ladder of Success

In January 1969, I was promoted to CDR. Like my previous promotions, it sort of snuck up on me. Buzz and I returned to D.C. late at night after a trip to Omaha trying to work a deal with SAC on their IROL requirements and I got the news walking out of the NMCC on my way home; one of the shift guys I knew yelled out to me 'congrats.' It was a big surprise in that I was unaware that the promotion board was meeting much less that I was in the zone. The next big step up the ladder occurred when Ned Humphrey due to retire in October 1969, recommended I take his place as 'the' DIA rep to the ICRS. Just as important to my career prospects was my new boss, retired Air Force Colonel Henry C. Howard. Mr. Howard was former head of the NRO Satellite Operations Center (SOC), future Chief of Station for the KH-11 ground station (Area 58), and my greatest boss ever.

Even though I was getting ahead, I was still a junior Commander and had an Air Force LtCol as my reporting senior. Looking at my fitness reports from that period, one wonders what impact one O-5 recommending another O-5 for accelerated promotion would have on a selection board. But this artifact was offset by a stream of reporting endorsements by other, more senior officials, as well as a review by a senior Navy Officer. I was also trying to get better in other ways as I completed both the VICORE and Evelyn Wood speed reading courses. In an even more career oriented enterprise, in 1968 I enrolled in the Graduate School at the University of Maryland and over the next two and a half years, I would toil at my studies and eventually be rewarded with a MA in Government and Politics. Most of my course attendance was at night (7-10 PM) which made for a long day but the grind was made fun when I would arrive home at 11 o'clock in the evening and Mary would greet me with a hug and a kiss, pastrami sandwich, dill pickle and a cold beer. There were a few courses I could not take at night so I just snuck off from work to spend half a day at College Park, not asking permission because I wasn't sure what the answer would be. Even though this period was at the height of the very unpopular Vietnam War (especially on campus), I made it a point to always wear my uniform to class, hopefully to disabuse my classmates of the so called 'military mind' fiction.

Tagboard

My fascination with aviation which began in a navy squadron, continued unabated as every so often I got involved with the airborne reconnaissance part of the NRP. During my first DIA tour, this association concerned a high performance, ram jet reconnaissance drone designated the D-21 but better known as Tagboard. Recognizing that the U-2 would soon be vulnerable to modern air defense systems, Lockheed Corp. began development of this high altitude Mach 3 reconnaissance drone in the late 1950s as an adjunct to the OXCART program. The A-12 Oxcart aircraft, which has a fascinating history of its own, was tested in Area 51 in the Nevada desert which gave rise to many of the alleged UFO sittings in this region. It had some operational success flying a number of Black Shield missions over North Vietnam and Korea in 1967/1968. However when it was deployed in theater to Kadena AFB, Okinowa, the

President did not authorize its use over China; this would not be the case for Tagboard. Oxcart and the D-21 drone were developed as part of the same program. The photo on the right taken during testing, shows the Tagboard drone mounted on the vertical stabilizer of a modified Oxcart, the M-21. Due to its ramjet propulsion system, the drone engine, which needed to be started at high speed and altitude, was ignited while still mated to the mother aircraft. Midway in the development stage, a disastrous separation during a flight test destroyed both the M-21 and D-21, killed the launch controller and resulted in an immediate redirection of the program. From that point on, propelling the ram jet to speed would be done by rocket assist, mounting the booster and drone on the wing of a specially configured B-52H.

When I became aware of the program, I quickly adopted it as my ward, built a briefing package and started on my rounds with senior Pentagon officials much like I did using high resolution satellite imagery at CINCLANT. I developed a following when, beginning in late 1969, operational missions over China started to be flown, but unfortunately none of these operational missions were successful. Two suffered loss of the film capsule during the recovery phase and two crashed or were shot down. The last mission in the spring of 1971 was monitored well into central China but never returned. The program was canceled shortly thereafter and my briefing efforts came to an ignominious end. What I did not know at the time was a candidate program manager for the project during its early stages was none other than my current boss, Mr. Henry C. Howard, at that time a Air Force LCOL. As far as the A-12 Oxcart aircraft was concerned, after a competition between the CIA and Air Force, it was retired in 1968 and for the next decade, its mission was assumed by its Air Force sister, the SR-71.

Appendix B contains excerpts from a 1967 Director NRO memo to the DepSecDef giving him an update on specific issues surrounding the Oxcart and Tagboard programs. There is also a marvelous book by Paul Crickmore on the A-12/D-21/SR-71 vehicles entitled *Lockheed Blackbird-Beyond the Secret Missions,* and on the main floor of the Boeing Museum of Flight in Seattle, you can see both an M-21 and D21. Another informative source is the web site at http://en.wikipedia.org/wiki/D-21_Tagboard.

ICRS and the SOC

I developed a close association with the NRO's Satellite Operations Center (SOC) principally through my dealing with the ICRS. The SOC was where the expertise on satellite systems performance resided and about which I wanted to be fully cognizant. A CIA guy, John Schadegg, who had a full time staff of CIA employees working for him, chaired ICRS. Committee membership included representatives from CIA, DIA, State, NSA and the Military Services; two or more reps from CIA and DIA attended most of the meetings, as the stakes were high in the prioritization battles being fought. The moderating forces were the ICRS Chairman and two NRO consultants from the SOC, Hal Bonney and Ben Niebur. These two sat in principally to insure that ICRS did not task the NRO to do the impossible. At the start, being a DIA third stringer, I listened carefully and learned some valuable lessons on how to succeed in this arena: never provide the NRO with requirements outside the collection system parameters; submit requirements that could easily be collected; and look for requirements that could be collected as a bonus when meeting some other agency's needs. I fully understood how to technically do this only because Hal and Ben put me under their wing and spent many hours educating me. I was encouraged to visit their spaces in the Pentagon basement and for the rest of my tour, I was a constant presence there.

And what a great group of guys were holed up in BD944. Till then I had thought submariners were the most elite organizational group but quickly changed my mind when getting to know the SOC 'space cadets.' Drawn from all the services and agencies (one I knew from my CINCLANT tour–CDR Tom Curry), they were as affable and forthcoming as could be. And the result was that I really got to know how the SOC functioned and the limitations of tasking satellites. Soon after I got to be the ICRS rep, I took a VIP trip to the West Coast as the NRO flew the entire ICRS to Edwards AFB for a U-2 familiarization (near vertical takeoff demonstration); Beale AFB to do the same for the SR-71; San Jose to look at, among other things, Lockheed's Shake & Bake chamber; and Vandenberg AFB to witness a Gambit satellite launch. Greatest field trip I ever took. The then head of the SOC, Col. Ed Sweeney, lived in Tantallon MD, a place I took note of and where I would live with my family four years down the road. After leaving government, I would regularly attend the annual 'Mushroom Factory/Friends of the SOC' luncheon in the DC area and renew my acquaintance with a great group of people.

Vietnam

As a rookie DIAXX staffer, I was given the task of monitoring reconnaissance related message traffic originating in the Vietnam theatre. I was reasonably familiar with what was going on from my experience onboard the Independence where I briefed on ship operations during its previous Westpac cruise. The Air Force was flying extensive tactical recce missions from South Vietnam or Thailand, and the Navy from aircraft carriers in the Gulf of Tonkin. This was in addition to A-12, SR-71 and various drone missions under national auspices. My job was to read all the incoming photo interpretation reports received each morning and pick out anything of potential interest to my superiors. Time consuming and not very rewarding work but useful in getting me to appreciate how complex fighting a war is in terms of sorting through the wealth of information available to commanders and how so little of it really matters, and more germane, how hard it is to find the gem/s that make a difference.

An example of the latter occurred in early 1968. Up to that time the North Vietnamese had not yet fielded an armored tank unit in South Vietnam, and it was considered impossible for them to do so without being spotted by aerial photo reconnaissance. Despite visual sightings of armored forces in the area, a detachment from the US 5th Special Forces Group and about 500 Montagnard irregulars at Lang Vei, relying on this imagery derived intelligence, were surprised when Soviet made PT-76 tanks attacked their camp with disastrous results for the defenders.

Not having experienced battle, it is hard for me to imagine how the Green Berets felt when the tanks suddenly appeared before them on February 6, 1968. Thirteen PT-76s, of the NVA 202nd Armored Regiment spearheaded the assault with the defenders fighting back with Recoilless Rifles and ineffective Light Anti-Tank Weapon. The photo on the right shows two of the PT-76 that were destroyed in the battle. The Lang Vei camp was eventually overrun, with the PT-76s using their turret-mounted spotlight to machine gun any irregulars who panicked and ran out of the underground bunkers. Despite a gigantic photo interpretation effort by the richest, most powerful nation on earth, we could not ferret out this critical morsel of information, i.e., enemy tanks in the area. The lack of good all-weather/night sensors was part of the problem but more importantly, the North Vietnamese were masters in hiding the movement of equipment south on the Ho Chi Minh Trail.

Occurring shortly after the Tet Offensive, this episode solidified the anti-war protest in this country, which was arguing that the US was supporting a corrupt regime against an indigenous revolt. The presence of Soviet PT-76 tanks should have disabused the protestors of this notion, as the Soviets produced some 12,000, supplying them liberally to client states at odds with the US in their effort to drain our country dry in the larger cold war effort. I disliked the protesters not only on these grounds but also because they made me leave for work at 5:30 in the morning to avoid their traffic-tying-up demonstrations outside the Pentagon.

I had one other significant connection to the war that I regret. In 1969 just after settling in as the DIA representative to the ICRS, I got a call from Col Franklin, one of the O-6s assigned to the Chairman JCS Staff Group, asking me to come to his office as the Chairman had something for me to do. When I got there, he told me that General Wheeler wanted to insure that no satellite imagery would be taken of southern Cambodia in the general vicinity of the Parrot's Beach, and to maintain this embargo until further notice. Thinking that this was a precaution to mask the marshalling of US forces preliminary to an encircling maneuver isolating North Vietnam forces in the south, I complied with the request and dutifully convinced my fellow ICRS members to eliminate that area as one of its collection requirements. It was not until the extent of our involvement in Cambodia surfaced in 1970 did I realize that my action was clearly one of covering up the bombing of what many considered a neutral nation. For those who have looked at imagery of areas carpet-bombed by B-52s, even in heavy foliated jungle, it has a signature unlike anything you have ever seen and something that could not be disguised or explained away. This bombing campaign was a prelude to the invasion of Cambodia and became the proximate cause of the Kent State protest and subsequent shootings. I look back at the event now not with regret for doing something wrong but for being naive and stupid about what goes on when fighting a political war. But it did have a salutary outcome; it made me more both critical and cynical about military involvement in wars of this type.

Near Real Time Imagery

Looking back at what eventually happened, there seems to be little doubt that a single product of mine was instrumental in both making my career and influencing the direction of the NRP. A twenty-page report, typed on my office Remington and illustrated with hand drawn graphics, entitled *Satisfaction of Military Operational Requirements by Near RealTime (NRT) Imagery* was initially read by only three people. But it precipitated a torrent of actions. To see why it had such an impact, it is necessary to look at the way the NRP was organized, especially how the government attempted to keep the rivalry between the Air Force and CIA within acceptable limits.

The NRP was created in 1961 and reorganized in 1965 when an Executive Committee (EXCOM) was formed. The EXCOM then consisted of the Director of Central Intelligence (DCI), the Deputy Secretary of Defense, and the Special Assistant to the President for Science and Technology. The committee was to recommend to the Secretary of Defense (SecDef) the "appropriate level of effort for the NRP, approve or modify the consolidated NRP and its budget, approve the allocation of responsibility and the corresponding funds for research and exploratory development for new systems." Specifically funds would be made available to the various program offices to meet their assigned sensor development responsibility by technology area. In 1969 DIA's role in the NRP was significantly enhanced as Mr. Hughes convinced the SecDef to seat the Director DIA as the Military Intelligence Advisor to the EXCOM.

Now with a viable and powerful decision mechanism in place, the deployment of forces for the big NRT battle began with skirmishes involving requirement definition. The first JCS paper I worked on was a memo to SecDef reiterating the need for 'real time imagery,' tracing the requirement back to the SAMOS era, an electronic read out competitor to the CORONA system, one that literally and figuratively never quite got off the ground. Shortly after this paper went across SecDef's desk, Col Lew Allen, then Director of the NRO staff stormed into Mr. Hughes conference room one day (John and I were sitting there going over his Quarterly Reconnaissance Briefing) and Allen growled out "Goddamn it John, don't send letters like this upstairs...if you want a capability developed, get the intelligence community to tell us what they want." (Allen eventually headed NSA, became Chief of Staff of the Air Force and after retirement ran the Jet Propulsion Laboratory). This exchange started a process that within a year resulted in the USIB approving the requirements for near real time imaging satellites. The general nature of the requirement was derived from a number of studies conducted in 1968/69; one produced by Roland Inlow, Chairman of the COMIREX, was by far the most thorough and convincing and showed the value of NRT Imagery for Indications & Warning, Crisis Management, Scientific and Technical Intelligence, Order of Battle, Arms Control Monitoring and Military Operational Support.

While these requirements were in the process of being formulated, the CIA and Air Force were each busy developing a satellite system that would satisfy them. The Air Force chose to modify an operational film return system by adding an on-board processor and data relay capability with a link back to the earth using existing ground stations; this system was identified as the FROG (Film Readout-Gambit). The CIA decided to push the technology envelope by using a brand new electro-optical (E-O) sensor, new collection and data relay satellites as well as a new ground processing and exploitation facility in the DC area. Thus the differences were large between the programs, both in technology and cost. The photo below is an artist's conception of the proposed CIA spacecraft in flight.

And while all of this was going on, I provided my recommendation on where DIA should come down on the competing proposals. Now ensconced in the back of the bowing alley with some privacy, I wrote a report that argued that the development of the E-O imagery satellite system as proposed by CIA would not meet the military operational support requirement. In the early 60s, I had taken the Military Studies Curriculum offered by the University of Maryland and had built a fair size military history library to include battlefield maps. Using this as reference material, I showed that the CIA 'postage stamp' sensor array did not have a robust enough capability to adequately cover military staging and maneuver areas. I brought the report to Mr. Howard, he moved it on to Mr. Hughes and in less than a week, it was up in front of LtGen Don Bennett, the Director DIA.

Gen Bennett as the Military Advisor to the EXCOM broached the problem in general terms to them and being a very prudent man, at the same time directed the Production Department of DIA to look at the my findings/scenarios to both verify and fully document them. CAPT Wendell Furnas was brought in to run the effort and in early 1971, it did indeed validate my findings. Now with a decision deadline approaching in June 1971, the ball was back in the court of the Air Force and CIA to respond. The Air Force proposal already had the necessary flexibility to operate in what was called the "stripping" mode but the CIA needed to go back and redesign their sensor array and that took some time. Just as I was departing DIA in mid 1971, the EXCOM decided that the Near Real Time imagery satellite system would be the Air Force's FROG.

As a result of my efforts on this project, I was awarded a Legion of Merit (medal is shown on the right) and I am very proud of it, since it is not normally given to O-5 staff officers. But even as the EXCOM decision was being made, the CIA was modifying the sensor array in their proposed system to add a new feature (the stripping mode) that would indeed meet the Military Operational Support requirement. And with this modification, the CIA showed the tremendous flexibility their proposed system had in terms of growth, something where the FROG had serious limitations.

Winding Down

At the end of my tour it became very apparent that the conduct of day-to-day DIAXX business was significantly affecting the large multidiscipline DIA Collection Directorate (DIACO) in an adverse way, and that consolidation of the two looked like a smart thing to do. I bring this up since my sole contribution to what eventually became the Directorate for Collection and Surveillance (DIACS) was in its naming. As one might imagine, the resultant office locations and titles were very important to the parties involved and negotiations on these matter between XX and CO took months.

But as all of this got worked out, no one had given any thought about the name of the new Directorate. And as mundane as one might thing such a matter was, it was treated as a crisis matter the night before the Director DIA was to announce the merger. Mr. Hughes, Mr. Howard and I sat around late into the evening trying to come up with a name that reflected oversight of all collection activities, but not use the old DIACO title; we wanted everyone to know that DIAXX had gobbled up DIACO, not the other way around. Not being able to leave the word "collection" totally out of the title, we arrived at the DIACS moniker.

One potentially meaningful end of tour project that affected the new directorate was something called Area Reconnaissance Specialists. I suggested that each officer in the newly created imagery division within DIACS be assigned specific geographic area/s of interest and be responsible for assessing the validity and timelines of the need for imagery as set forth in the IROL, and more importantly, be willing (and competent) to stand up for it under attack. This was not an area under my cognizance but I felt very sure of myself in proposing it, and had some confidence that Mr. Howard, the new chief of the division, would buy off on it. The opposition was mostly one of not being invented here, and while it eventually was approved and implemented, it never got the enthusiastic support it needed and eventually was abandoned.

My very last big project was one that became a standing joke with my colleagues. It started simply enough when the Deputy Secretary of Defense (DepSecDef) asked the DNRO and the Director DIA to jointly develop some charts that showed where money was spent in the imagery sector of the NRP and how this related to the resultant intelligence product. The then DepSecDef, David Packard of HP fame, was a very savvy guy and simply wanted to know what value we were getting for the vast amount of money we spent on satellite reconnaissance. I was never sure why the project gravitated down to me other than being the current boy wonder and having a reputation of being able to do the impossible (which this project was). But I jumped in and produced what became known as the 'Packard Pie Charts.' On the surface, they looked good showing costs and related intelligence benefits in very broad terms, something one might look at and see a meaningful relationship. But anyone familiar with the subject would recognize them as froth. I never found out what Packard or other people on high thought of them, and while they did seem to answer the mail, I was mercilessly ribbed about it. I still chuckle at the cartoon board shown below, presented to me at my going away party by the people I worked with. It captures perfectly three years of a wonderful assignment in the form of two pie charts.

Afterward

In the introduction, I mentioned how much I owed many of my bosses, some who really went out of their way to help me. Nowhere was this the case more than in DIA. First was Mr. Hughes; I initially came to his attention at a DIA picnic where his kids and mine got together to play softball. Somehow or another he sensed a kindred spirit in me, and shortly thereafter, he came over to my cubby-hole to chat. Then he began to special task me and after a while, asked me over to his office to talk about various issues he was involved in. We would share personal items, e.g. how our wives (his was Paula) coped with our late hours, and while we never really became really close friends, we developed a good mentor/protégé relationship. I would latch on with Mr. Hughes twice more and each time he would move me along nicely in my career path. The final time was when he became the Deputy Director for Defense Intelligence and hired me in 1979 as his Special Assistant, a job I held for three years as a civilian GS-15.

As for Mr. Howard, I can remember talking to him about his family only once and our personal relationship was never close. We did however become professionally bound together; one night about seven as I was leaving the office, I peeked in and saw him at his desk with a worried look that said he had a big problem. I asked "can I help" and after some hesitation, he looked up at me and said "yes" and after that we off to races as professional confidants. Reading his fitness reports on me, you can see that in many ways he regarded me as a favorite son. Our families did interact once as his wife Burma approved a serious transaction between the Howard and Bausch households, the gift of their ancestral piano to us, a large upright that we had for almost 30 years (it was donated to the Harmony Hall Recreation Center in Fort Washington upon our move to Caswell Beach). He later went to work for the CIA as the Chief of Station at the KH-11 Ground Station and we kept in touch over the years until his death in 2006.

I left DIA in early June 1971 for my next duty station at Lowry AFB just east of Denver to become the Officer in Charge (OinC) of the Naval Unit and Deputy Director, Armed Forces Air Intelligence Training Center (AFAITC). Back in Washington, Dr. Land of Polaroid fame got involved in the near real time imagery debate and convinced President Nixon to reverse the EXCOM decision and approve the CIA plan to build what would eventually became the KH-11 imaging satellite. In November 1971, Henry Kissinger announced this decision in a memo to all concerned and five years later, the first KH-11 rocketed into space. For an entertaining read on this and a bunch of other historic Air Force/CIA battles, check out the *Wizards of Langley* by Jeffrey Richelson. In another piece of unfinished business, I had one more class to take for my degree at Maryland which I did by correspondence. In August 1971 I received my Masters of Arts diploma in Denver by mail.

Chapter 7-Armed Forces Air Intelligence Training Center (1971-1973)

Reporting Aboard

The Bausch family arrived in Denver mid-June 1971 and thanks to CDR Vic Burch, OinC of the Naval Unit, we temporarily got two suites on Lowry AFB in the VOQ, next to the officers club which had a swimming pool for the kids and just down the street from my new office. The first order of business was the change of command ceremony that took place on the tarmac of the old airstrip with my family in attendance. It was a hot day and we had some keel-overs, not a nice experience for my kids to witness. One of them later said it seemed like we were torturing the lined up sailors. I got to troop the formation and while it was not all that great a day, you could never tell from the photo on the right. It did not take us too long to find a nice place to live south of the base on East Colorado Drive in Aurora, which allowed me to begin assessing my situation and I decided to set some goals and objectives . Lowry was a large base, training some 15,000 students a year, 2,000 of which were Navy and Marine Corps. Besides air and photo intelligence courses taught at AFAITC, instruction for Navy and Marine Corps students was provided in medical equipment, precision measurement, aerial photography and logistics. The permanent Navy staff, mostly instructors in the schools, numbered around 50.

Within months I had fairly clear in my mind two things I wanted to do: one was to change the Officer in Charge designation into a Commanding Officer, and the second was to either move the navy intelligence training out of Lowry to a fleet environment, e.g. San Diego or Norfolk, or failing that, reorient the Navy officer course to focus more on naval operations and less on air intelligence. To effect both these courses of action, I would need a lot of support from my bosses who were many. The Commandant of the Ninth Naval District in Great Lakes, North Chicago, pretty much left me alone until just before my tour ended when a serious military justice problem arose (more about this later). My AFAITC boss, LtCol Jim Manley, treated me as an equal but we had a difficult chain of command reporting to two Colonels and a Major General, all sticklers for the Air Training Command way of doing things. But to everyone's surprise but my own, I liked the way they did things, and fell in line easily. Most of the Navy instructors couldn't stand the documentation requirements but I generally thought it made you think a lot about what you were trying to do before doing it. Plus my outlook on the Air Force way of doing things improved greatly when the Legion of Merit award came through and MGEN Samuel, Training Center Commander, pinned it on at a base-wide awards ceremony (photo on right). My third boss was an administrative one in Memphis TN, RADM Valdemar Lambert, CNATECHTRA (Chief of Naval Technical Training). For some odd reason both he and his boss, VADM Malcolm Cagle, Chief of Naval Training took a keen interest in the Navy Unit and as a result I regularly interacted with them.

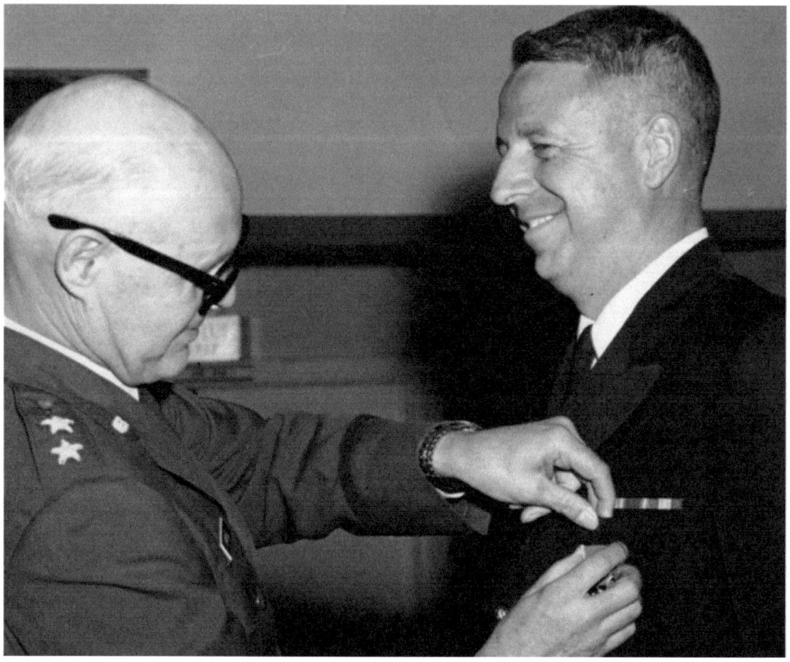

They visited Denver every once in a while and I would regularly travel to Memphis to report to them. Prior contacts helped here as both my skippers on the Independence were now a part of the training world: John Fox as Chief of Staff for Admiral Cagle and Blackie Kennedy as a hatchet man for Admiral Lambert. And finally, my fourth boss, the Director of Naval Intelligence, at that time RADM Rex Rectanus, while he did not write a fitness report on me, had to approve moving out of Lowry or changing the course curriculum, and from a personal standpoint, would need to buy off where I would be next assigned. Or so I thought.

Naval Unit Commanding Officer

The decision to pursue changing the Naval Unit status to a 'Command' was self-serving to a large degree. A previous request had been turned down ostensibly due to a lack of a host-tenant agreement between the Naval Unit and the Air Force and I could very much have left it there. But I wanted to be a 'Commanding Officer.' So I reopened contact with CDR Frank Picher in the office of the Director of Naval Administration, and over the next year we worked the problem until in November 1972, I got what I wanted. My arguments were based on a number of matters related to discipline, billeting, welfare and recreation but mostly image. Lowry was the pre-retirement tour for many Air Force officers; at that time there were 25 Colonels and 50 LtCols on the base, pretty much all of whom were senior to me. In one of my letters to

Washington I wrote "My authority is equal to that of an Air Force 2nd Lieutenant, who functions as the Commanding Officer of the Student Squadron, which is somewhat equivalent to our naval unit." I think this was the only compelling argument and eventually the one that won out. The only visible sign of the change was my title on official correspondence, and the Commanding Officer (Shore) breast insignia I was then authorized to wear on my uniform.

But I would note here that being a Commanding Officer is as large a responsibility as one can have in life. Navy Regulations Section 700.802 reads "The responsibility of the commanding officer for his or her command is **absolute**..." Thus no matter what directives, instructions, orders, etc. from higher command might have you do, the Commanding Officer is fully responsible for the success, safety and well being of his command and nothing can absolve him of that. So to do the job right you need to be a manager, judge, jailer, quasi spouse or parent, confidant, advisor and whatever else it takes. "Skipper, my husband didn't come home last night, what should I do?" is not something you brush off by saying " go talk to the Chaplain?" If ever the buck stops somewhere, it is in the lap of a US Navy 'Commanding Officer.'

Air Intelligence versus Naval Intelligence

As for my other objective, inasmuch as I was a Deputy Director with no real responsibilities other than backing up Jim Manley, I spent the first few months at AFAITC attending classes and trying to get a sense of what the center did training-wise to make a better air intelligence officer. I already had a bias against what I considered narrow air intelligence training and nothing I initially observed at Lowry changed my mind. Our navy officer students had experienced almost nothing in their lives but school and knew little about the real navy. So my first preference was to move them to a location where they could at least go aboard a ship or some fleet intelligence activity and see what actually was going on navy-wise. I broached this idea at the August 1971 Naval Intelligence Training Conference in Wash DC and it was coolly received. The old hands felt that this was not the time to overturn the October 1962 DOD directive that consolidated air intelligence training under Air Force aegis.

So I went back to the drawing board and dreamed up the idea of revising the course for the navy officer students to focus more on intelligence support to naval operations rather than just air intelligence. I let my Memphis bosses know what I was doing in an August letter where I pointed out that " specialized air intelligence training is out of line with the mainstream of the Naval training atmosphere." And then in conjunction with my AFAITC navy officers, in September 1971 at the Air Intelligence Conference at Lowry, we surfaced this idea to the world. In general, there was tacit support with the implied "let's see what you can come up with in the way of a specific course curriculum." With that foot in the door, I turned my guys loose on the problem and within months they had a curriculum document in hand, and produced a set of briefing slides for senior officials focused on closing the deal. The new curriculum was a big departure from the existing 22 week joint AI/PI curriculum in that almost 80% of the classes were new to include seminars, multiple exercises and field trips to naval operational facilities. Next we got the Air Force to provide a plane for a December trip to the east coast, stopping at Wash D.C., Norfolk VA, Albany GA and Jacksonville FL to brief the most affected navy commands. There were six of us on the trip and it was a success, due in large part to having each navy instructor solidly behind the new course. We followed this up with a similar trip in January to the west coast, visiting San Diego and the Bay Area and again met with enthusiastic support.

And then recognizing there was probably no one out there exercising veto power, we began to internalize the new course in the AFAITC curriculum. The first Naval Intelligence Officer class (NIOC) began on 26 July 1972. We made some modifications as a result of hands-on learning experience, changed the lesson plans accordingly and with little fanfare, the final plan of instruction (POI) was approved in January 1973. The sample page of this 108 page document below shows the kind of change involved as the curriculum would now spend two full days on US and Soviet Submarine Forces.

PLAN OF INSTRUCTION — COURSE TITLE: Navy Intelligence Officer

BLOCK TITLE: Navy Undersea Warfare

UNITS OF INSTRUCTION AND CRITERION OBJECTIVES (1)	DURATION (HOURS) (2)	SUPPORT MATERIALS AND GUIDANCE (3)
1. U.S. Navy Submarine Force Without reference, diagram the composition and organization of the U.S. submarine force, and identify the major factors which make submarines an effective element of strategic and tactical warfare.	3	Column 1 Reference: 1 — CTS Reference: 3a, 4a, 4d, 71 Instructional Materials None Audio Visual Aids Selected Transparencies Selected 16mm Films Training Equipment Overhead Projector 16mm Projector Training Methods Discussion (3 hrs)
2. Soviet Submarine Force a. Without reference, diagram the current composition and organization of the Soviet submarine force, and analyze submarine employment as part of current Soviet naval operations, noting future development trends. b. Given selected imagery and imagery interpretation equipment, locate and identify submarines, submarine bases, and associated facilities.	6	Column 1 Reference: 2a, 2b — CTS Reference: 3a, 5a, 5d; 12c, 14f (1)--(7) Instructional Materials Soviet Submarine Force Audio Visual Aids Selected Transparencies Selected Films Selected Imagery Training Equipment Overhead Projector 16mm Projector

PLAN OF INSTRUCTION NO. 30BR8051-4 | DATE 15 January 1973 | BLOCK NO. V | PAGE NO. 45

U.S. Government Printing Office: 1972 — 779 108

Unfortunately however there was a negative to the change, as it slowly eroded Air Force/Navy interaction and camaraderie at the center, something that would be become vital in the future for many of the students as they entered joint service command environments. Looking back, I also see a lot of irony about instituting the new course, i.e., a technical training center run by the Air Force providing mostly non-technical, operational intelligence training to Naval officers.

Reserve Officer Training

AFAITC also regularly conducted training for Naval Reserve Officers. They would show up for two weeks, with the predominant bulk of attendees coming during ski season. I had previous experience with reserves in the late 60s at NAS Alameda, the highlight being the Thursday night parties they threw for the PFAITC staff. They would often kid us that they choose Thursday night so as to experience the inevitable hang-over on a school day, rather than the more accepted rationale of providing them an early Friday getaway opportunity to get back home. One of our favorite repeat students at Alameda was a Braniff Airlines executive who would insure that the all stewardesses in the Bay Area at the time attended these Thursday night parties.

When at Alameda, I never thought much about the training we gave reservists; they were sort of vacationers and I had no experience with them on active duty. During the Cuban Missile Crisis however, Navy Reservists were called to active duty at the Naval Photo Interpretation Center to help us. For the most part, they showed up with little idea of where they could fit in the NAVPIC work structure. I had one assigned to my group, a Commander, who was totally unfamiliar with the organizations and processes we were involved with. Because we were so consumed with just keeping our heads above water at the time, there was little opportunity to educate him on the fly, and his prime contribution to our effort was making coffee and running errands

By the time I got to Lowry, there was some recognition in the naval intelligence reserve community that we needed to do better in terms of preparing then for their mobilization billets. When they showed up for training, we would know for example what their billet was going to be, but other than that, we had little power to tailor our instruction or recommend a reassignment to a more suitable job. So I saw the whole thing as 'more of the same.' Thus I mostly railed against the system and to the degree I could, involved myself in other potentially more productive endeavors. And I could do so in clear conscience since I had a LCDR on the staff whose primary duty was 'Reserve Coordinator.' Ben Mater, a good friend and former shipmate at PFAITC did a great job with this ball of worms. He did such a good job in fact that I recommended he replace me as CO Naval Unit/Deputy Director AFAITC when I received orders to leave Lowry.

The photo on the right shows me promoting him to Commander. Unfortunately as far as taking my job, a Captain who was ending his tour as a Naval Attache and wanted to retire in Denver, used his influence and got it instead. Note the uniform I have on in the photo-Winter Working Blue-which I authorized the command to wear in the cold weather season, something which my Admin Officer thought we should check with higher command. Knowing my four bosses all had different ideas about how the Navy Unit should be run, I never inquired what they were, and pretty much decided on my own what made sense.

Welfare and Recreation

While student hours on the day shift were straight through 0600-1200, instructors were expected to hang in to 1430. But I often encouraged them to leave after lunch to avail themselves of productive off base activities. I went so far that in the early spring of 1973, I declared 'Navy Ski Day' and gave the navy staff and students the day off to slush around Keystone and drink hot spiced wine. Skiing as one might expect was the one really unique bonus of being in Denver; all the reserve intelligence classes scheduled in the winter were fully subscribed to and the VOQ full of visiting navy brass that needed to check on how the Naval Unit was faring. Our family's favorite ski spot was Arapahoe Basin (shown in photo on the right) that was well up

the mountain on RT 6 (the Eisenhower Tunnel had not yet opened). It was quite cheap to ski then, as lift tickets from Air Force Special Services cost $5. But there were a lot of other activities that made the tour fun. During the warm months of the year, we had regular softball games: officer versus enlisted and Air Force versus Navy. A keg of beer (not having to pay for it) was usually the prize for the winner and plastic beer cups were scattered all over the field as the game proceeded. On other days, lunch time bowling was a favorite, with league bowling at night a fun time for both staff and their wives.

When snow was off the ground, golfing was also a big thing. The base league was very competitive with 45 teams competing. CDR Bill Halloran, Chief of the Operational Intelligence Branch, put together a great Navy team but had to raid the student ranks to do so. ENS Rich Barkell, who could smash a five iron further than any of his teammates could hit a driver, was the difference as the Navy Unit won the 1972 base championship. The photo below shows the team with their trophies (left to right, Bill Halloran, myself, Bud Humphrey, Dave Peterson, Larry Jensen and Rich Barkell).

And as the senior naval officer on the base, I was appointed to the Officers Club Board of Governors and absolutely loved the long business meetings over cocktails discussing how to improve club service. My wife and I also got to spend some quality time together. I fondly remember our forays to the Havana Inn for martinis at the bar while we waited to get seated for our favorite, Taco Burritos. Such nights out served as a marvelous romance re-kindler. But before you get the idea that Lowry was all fun and games, below is my sign off in a Jan 1972 letter report to Memphis. "That about concludes what little has happened here. Other than minus 15° temperatures, heavy snow and 100 plus mile per hour winds, it's a delightful place to be."

Professional Highlights

In April 1972 Jim Manley retired and COL John Oberg, former Assistant to the Director DIA, replaced him. One of the first things I did when John settled in was convince him to move my office into the training facility, something that offered me a lot more freedom to interact with the instructional staff. I have already mentioned the navy instructors and what a great bunch they were. Most had just come back from a Vietnam tour and I once said at a staff meeting that if I had to go into battle, they would be the ones I would want with me. While I was at Lowry, our instructors had a 100% promotion rate and a 100% approval for change to the 1630 designator. In other respects the Navy Unit excelled. The Military Wife of the year for 1971 was the spouse of one of our yeoman petty officers and in October 1972 we put on a show to celebrate the Navy's 197th Birthday. Admiral Clarey, Commander in Chief, Pacific Fleet (photo below) presented awards to several Navy personnel and then reviewed the Naval Unit in parade as the Air Force Academy Band played Anchors Aweigh and A-4 Skyhawks from VA-127 based at NAS Lemoore CA streaked overhead.

In addition to coordinating everything, our PIO, LT Les Joslin made sure that the event was shown in great detail that night on the local TV outlets. Another interesting inter-service interaction was the Navy Unit attendance at the 1972 football game between the Air Force and Naval Academies in Colorado Springs. We got the Lowry transportation people to get the Navy attendees back and forth in an Air Force bus and we sat in full uniform, seated with our dependents in the Air Force rooting section. What made this extra special is that Navy won 21 -17 in what was considered a big upset. Les Joslin also helped my career in another way. In his capacity as PIO, I asked him to put together a brochure on 'The Navy at Lowry' and make it slick as well as informative. He did a beautiful job and put out two issues (January 1972 and 1973), which I had copies sent to everyone and his mother-in-law. Just to make sure I didn't get too big of a head, my former associates at the SOC returned a copy of the forwarding page of the 1972 version stamped as a codeword document (see below). Note the comment in the upper left by Hal Bonney about the pie charts coming back to haunt him.

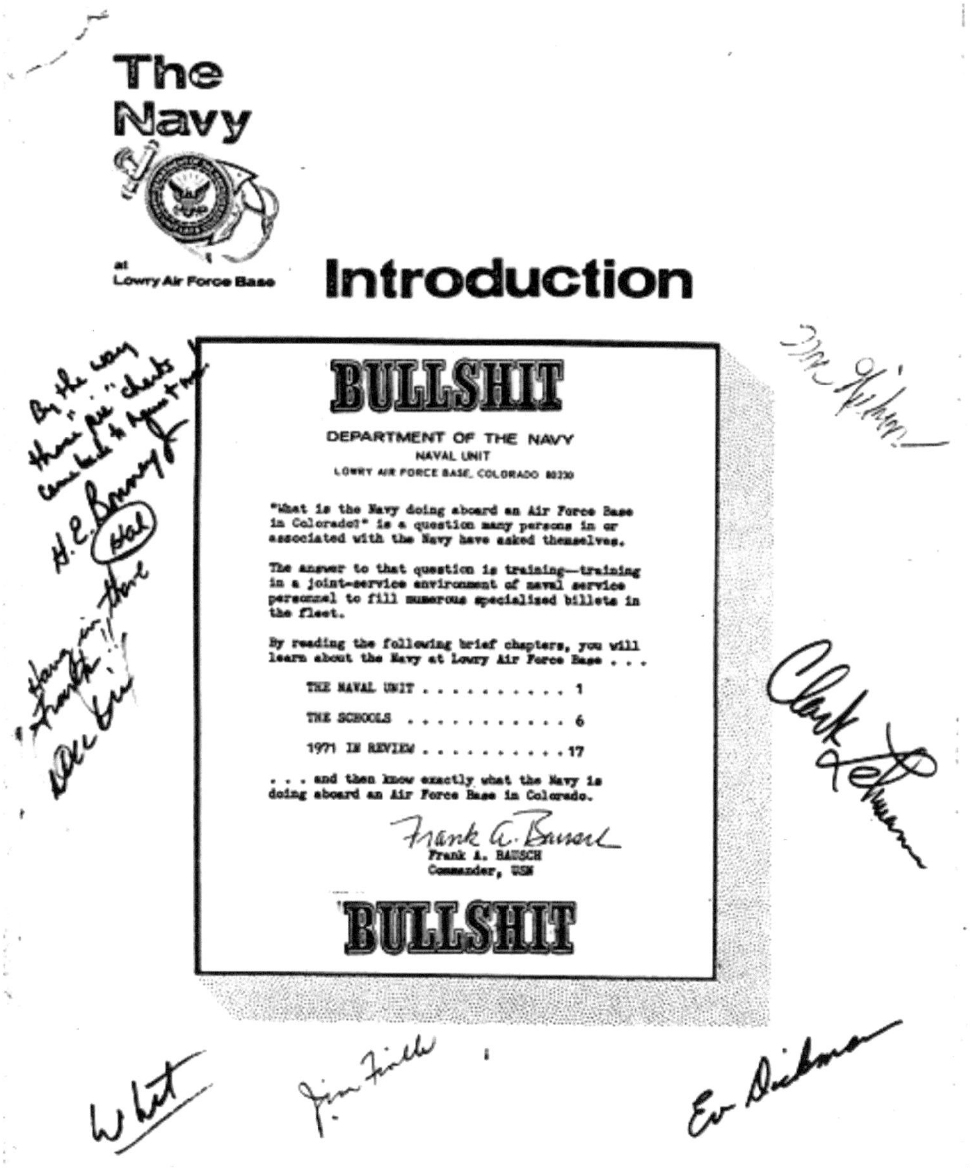
The Navy at Lowry Air Force Base

Introduction

BULLSHIT

DEPARTMENT OF THE NAVY
NAVAL UNIT
LOWRY AIR FORCE BASE, COLORADO 80230

"What is the Navy doing aboard an Air Force Base in Colorado?" is a question many persons in or associated with the Navy have asked themselves.

The answer to that question is training—training in a joint-service environment of naval service personnel to fill numerous specialized billets in the fleet.

By reading the following brief chapters, you will learn about the Navy at Lowry Air Force Base . . .

THE NAVAL UNIT 1

THE SCHOOLS 6

1971 IN REVIEW 17

. . . and then know exactly what the Navy is doing aboard an Air Force Base in Colorado.

Frank A. BAUSCH
Commander, USN

BULLSHIT

While the brochure was useful in a general sense to potential users, it was also a promotional vehicle to move my career along, something I started to think seriously about. I also took the step of reviewing my fitness reports and noted that those coming out of Memphis had CNATECTRA ranking me with officers on his staff. I wrote both RADM Lambert and his successor RADM Bergner, asking them to revise their past reports and one actually did. I think it was clear to me about then that I had a chance for promotion to Captain and while this didn't affect work, I did try to become more aware of the niceties in dealing with my superiors than I had previously.

I also had another brush with academia. In early 1972 I applied to the University of Denver for admission to its International Studies PhD program and after some intense interviewing, was accepted (see Appendix B). I then twice failed to get approval from the Navy to take nine months off to attend required on-campus courses; the rest of the PhD work could have been done in absentia. So I had to inform the university that I could not accept their offer. The program I missed out on was a good one with one notable graduate, former Secretary of State Condoleezza Rice. Other than the credentials associated with the degree, I would have liked the opportunity to write a book. Recently I read *Dereliction of Duty* by H. R. McMaster, a masterpiece of military/political history about the Vietnam War that started as a University of North Carolina PhD dissertation. Returning from combat in the first Gulf War, then MAJ McMaster wrote his brilliantly researched book while in his 30s. I would have loved having the chance to do something similar.

But losing this opportunity worked out okay as shortly thereafter I got bumped up the command chain when John Oberg was temporary reassigned to Operation Homecoming (a rehabilitation effort involving newly released Vietnam POWs). For much of 1973, I got to play top dog as Acting Director, AFAITC. An Air Force O-6 from one of the schools on base could have been transferred in but the newly arrived Training Center Commander, MGen Alton Slay, seemed to like me and my background in reconnaissance (Slay already had a back seat ride in an SR-71). This together with some hangover from the Legion of Merit award probably swayed his decision. As an aside, Slay, who eventually would get his fourth star heading the Air Force Systems Command, was being stashed at Lowry AFB to get him out of the field of fire aimed at GEN John Lavelle for his unauthorized approval of protective reaction strikes in Vietnam. Gen Slay at that time was Lavelle's Deputy Chief of Staff Operations.

Trouble in Paradise

Reading all of this, one might think that, other than the weather, Lowry was a bed of roses. But that was surely not the case as the six cryptic tales of woe which follow, none of which I handled particularly well, will attest to.

My biggest regret is how I dealt with the revised Navy fitness report guidance issued in 1972 which was well intentioned but unevenly implemented. Trying to avoid a clustering phenomenon—one contemporary study noted that 90% of all officers were ranked in the top 10%—all officers within a specific grade were to be rank ordered from 1 to n. I complied with the expected disastrous results. A top notch officer who won a Bronze Star in Vietnam, got ranked last of thirteen officers in his grade because of 'less then exceptional instructional skills'; you can imagine what that did for his career advancement prospects. I should have just ignored the new guidance like so many other reporting seniors did, but alas I did not and feel very bad about it.

Women started showing up as students in 1972; that year we had the first woman naval intelligence officer successfully complete training at the Center. While the officers were not a problem, this was not the case with some of the young enlisted female trainees. I vividly remember one instance where an 18 year old student asked to see me and for the first few minutes held herself together as she asked me to help her get out of the Navy. And then the tears came: she didn't understand what she was getting into when she enlisted; she was lonely, missed her family and all she wanted to do was go home. What a feeling of helplessness as I told her there was nothing I could do. Considered an adult and not the teenager she actually was, as far as the navy was concerned she had made her bed and now had to sleep in it.

And then there was Martinelli, a navy airman who had fraudulently enlisted by lying that he had never performed a homosexual act but subsequently admitted doing so in a separate investigation. While processing him for an administrative discharge, he solicited sex on base from an on-duty Air Force Security Policeman. This really exacerbated the problem but to get rid of it quickly, I decided to take a chance and continue with the admin procedure, not bringing charges under the UMCJ. But before I got to that point, Martinelli changed his mind and decided to fight any discharge at all. And unbeknownst to me, he had a friend in high places, Sen Clairborne Pell of RI. Next came ACLU involvement, a civilian attorney and a complaint lodged against me. It was all resolved after I had left Lowry but not before I sweated a few bullets.

And if the Martinelli situation didn't give me a near permanent headache, the situation that evolved out of my inept handling of our first black navy intelligence officer student certainly did. Washington had alerted me when he came on board, and I was told to be sensitive (read helpful) about getting him through school. A good student in that he was conscientious and worked hard, he was a terribly slow learner. Through a prodigious effort of the part of the instructional staff, he successfully demonstrated all the required performance and knowledge skills and graduated. Where I went wrong was stating in his fitness report that I did not think he was qualified for all intelligence jobs and recommended his assignment to a Fleet Intelligence Center type environment. What I wanted to do was avoid having the detailer sending him out to an attack squadron where he might get involved in putting together an Alpha Strike. Turns out he took umbrage at the report which resulted in a rebuttal letter, a favorable endorsement by his next reporting senior, and a matter needing my attention long after I left Lowry. I never backed off and would do the same thing again today, but look for another way to go on record about the matter other than putting my recommendation in a fitness report.

Another problem was my handling of a drug case where three enlisted students were caught dealing LSD. It was subsequently determined that one was a 'drug kingpin' (we never did figure out why he joined the Navy or how he passed the screening to get into the intelligence business). I recommended he be given a general court-martial with a view towards a dishonorable discharge and prison time. This was something only the Commandant of the Ninth Naval District could order and his legal staff didn't share my hard line view on this matter. We went back and forth until a plea bargain compromise was reached but not before some heated phone calls between Great Lakes and Denver.

To top all of this, while Acting Director of the AFAITC, I took it upon myself to fire an Air Force civil servant, the resource manager for the center. People joke about the impossibility of doing this kind of thing but I genuinely thought I could. I built a good case and pursued it vigorously. But you cannot imagine the number of phone calls and visits I got from the base hierarchy asking me to back off and reconsider. Eventually I did when the miscreant agreed to retire, making the whole thing OBE. But I have my doubts whether I could have pulled it off given the waffling type support by my seniors. So it may be that firing a civil servant is indeed impossible.

Moving On

Despite all the turmoil this strange assortment of problems caused, on his return John Oberg recommended me for a Meritorious Service Medal (MSM) as an end of tour reward, something that went through without a hitch. While I was looking at potential future duty stations, my old bosses in the Pentagon were scheming to have me return there. Although I never saw it, I was later told that VADM Vince DePoix, the then Director DIA, sent a letter to the Chief of Naval Personnel specifically requesting I be re-assigned to DIA to work on a 'vital national security project.' And with that kind of shenanigan going on, any hopes I had to get the MSM pinned on in a Navy environment all but disappeared.

Chapter 8–Defense Intelligence Agency (1973-1976)

Getting Settled In

Although I knew I was going back to my old job in the Pentagon, the message orders only arrived in mid-June with instructions to detach that same month. Up to then I made no serious attempt to sell our house; I had seen too many situations where orders did not come through as promised and complications ensued. So on the last day of June, I left Denver sans family in our second car. Back in DC, I got a room at the BOQ in Cheltenham MD and spent all my spare time looking for a house to buy. Luckily, our Aurora home sold in late July; I flew back, loaded up the wife, six kids, dog and cat into our VW van and drove back to D.C. in three days. We settled temporarily in an apartment in Arlington VA and then got a big break finding our dream house under construction in Tantallon, MD. We signed the contract in mid-August, quickly rented a place five blocks away, where we lived until the house was completed in January 1974.

Return to the Pentagon

I reported to work the second week of July 1973. The re-organization two years earlier where the Special Activities Office had absorbed the entire Collection Directorate had now been fully realized. Mr. Hughes was now in charge rather than a special assistant and managed a large organization with broad responsibilities. This resulted in some major shuffling of spaces where the majority of imagery division offices were now downstairs in the Pentagon's mezzanine basement. Although nobody worked for me, I was given a title: Chief, Systems Development Section, a part of the Systems Branch which was run by an Air Force Col who had an Army COL selectee as his deputy. All of this was in the Imagery Division headed by my old boss, Mr. Howard. I got a desk in a small box sized space with portable room partitions separating the four of us who worked there. In such small confines, visitor and phone conversations were shared by all, and everyone benefited from that. We had one overworked secretary, Becky, who recognized our strangely stressed work environment, took pity on us, and helped us any way she could. I continued my habit of bypassing the chain of command as the situation demanded, spending as much time with my big bosses, Messrs. Howard and Hughes as was necessary. Neither of my two immediate superiors complained and I did try to keep them abreast of what I was doing. The only down side of the situation was the trek through the maze of corridors and security barriers to get up to the second floor in response to a 'need you up here right now!'

It was not long before the Meritorious Service Medal award from Lowry showed up at the Pentagon and Mr. Hughes presented it (see photo below) in a ceremony conducted in the briefing theatre with my family, including my mother, in attendance.

A New Role Model–Pete Doerr

My visits upstairs regularly brought me into contact with a new face, a real pistol named Pete Doerr. A Navy Commander who had an office adjacent to Mr. Hughes, he served as John's staff in reviewing the National Underwater Reconnaissance Program (NURP). As you might surmise, this program was a joint CIA/Navy endeavor, and similar to the NRP, an attempt to organize peace in the bureaucratic war over US underwater reconnaissance operations. Because of the high risks and costs involved in its execution, approval of prospective undersea missions required highly professional, unbiased judgments. Mr. Hughes's reputation as a straight shooter got him the job as the Defense Department voice in these matters and Pete was assigned by the Navy to help him do it. A blue water destroyer sailor, Pete was a designated Naval Intelligence sub-specialist and undoubtedly picked for the job because of his no nonsense approach to work. Stories about him abounded, many about confronting some Admiral or other senior official and dressing him down for some misdeed. Pete had a dour personality, very curt and unforgiving of mistakes but I immediately took a liking to him and in effect he became a real life role model replacing the remote, historical figure of Joe Stilwell. Maybe because his wife's name was Mary and he had a bunch of kids had something to do with it. The image I remember most of Pete was him sitting in his office at his perfectly clear desk, and I mean perfectly clear (no in-box, pen set, etc.), doing the Washington Post crossword puzzle in pen. It never bothered him that projecting such an image might do some harm to his career. And the first time I heard the word "shibboleth" was Pete arguing against the allegedly 'unbiased views' of the CIA. I keep up a correspondence with him after he left DIA when he assumed command of a destroyer squadron. He retired as a Captain, set down in Cape Cod and we kept in touch until he died in 1992, the same year as Mr. Hughes.

Dissemination Options

As I settled into my job, I soon recognized that there were a lot of high expectations about what I was about to do. In the two years I had been away, the JCS had approved and forwarded to the Secretary of Defense, recommendations about near real time (NRT) imagery that proposed four new initiatives:

- a technology project to demonstrate wide bandwidth communications needed for NRT imagery.
- a facility to be located in the National Military Command Center (NMCC) to coordinate the collection of NRT imagery with other reconnaissance systems.
- development of equipment for tactical users to handle NRT imagery .
- a distribution mechanism to rapidly move NRT imagery from the Washington D.C ground station to theater command headquarters.

The Chairman of the JCS formed a Joint Staff/Services NRT Task Force under the Director DIA to implement this directive and the Director made me its first and as it turned out, only chairman. I knew most of the assigned service and joint staff representatives and we quickly agreed upon a division of labor. The wide band demonstration project would be run by a sharp DIA Air Force Major (Tom McGraw), whose history included the misfortune of seeing a cherished space experiment he worked on for two years fail during deployment. Tom was one of the four officers I shared office space with so I easily kept abreast of how work proceeded on his part of the project. The photo on the right shows the Chairman of the JCS, Admiral Thomas Moorer accompanied by Tom, inspecting the test rig on the Pentagon roof. Connecting the National Photo Interpretation located in the Navy Yard with the NMCC, the Raytheon built equipment (AN-GRC 173) provided a 240 Megabit per second capacity using 37 GHz millimeter wave technology to rapidly transmit imagery to military decisions makers in the Pentagon.

The development of the Collection Coordination Facility required technical expertise as well as seniority to implement, and we got a winner, COL Chris Millson from the NRO, to head the effort. The tactical dissemination project was given to the Army who already were working this problem. Each of the services had at that time, in one form or another, TENCAP (Tactical Exploitation of National Capabilities) Offices. All were manned by sharp guys who were moving along smartly and I was not about to interfere; I helped when and where I could. My part of the action was getting NRT imagery out of town into the hands of four Unified and Specified (U&S) Commanders: Strategic Air Command in Omaha, European Command in Stuttgart, Germany, Pacific Command in Hawaii and the Atlantic Command in Norfolk VA.

Halfway into the project I had second thoughts about whether these commanders really needed imagery in near real time or whether they could live with imagery derived information, a method that conserved bandwidth by sending only the target signature/changes to an imagery data base already resident at their headquarters. This would be something very much like what we did with Air Target Material (ATM) charts in the late 1950s/early 60s using U-2 and KH-4 imagery. I talked about this with Mr. Howard who sensed my ambivalence and asked me whether I was getting cold feet. I said "no", and went on with getting the project realized. Looking back now, I think I made a mistake but in any case, the JCS in March 1974 approved my Dissemination Options Paper that recommended SECDEF approve the transmission of national reconnaissance program imagery to the four headquarters. The implementation scheme to do this was known as the Defense Dissemination System.

In the JCS staffing process, white colored papers called flimsies, are drafts being circulated at the service or agency action officer level; buff color papers have gotten to the next step which requires coordination at the planners (O-6) level. Papers surviving this step become green for consideration by senior (three star) level officers in each service. Finally, a red stripe on the green page indicates approval by the Joint Chiefs. An excruciating process for the action officer wanting to get something done but because of it rigor and length, a red striped paper represents a truly authoritative decision. The 'Dissemination Options' red stripe decision sheet is shown in Appendix B.

Imagery Mix/Search Studies

One of the interesting things about working on NRP related problems was the interagency participation, something I very much enjoyed. As my previous experience with the ICRS was very positive, I looked forward to these study groups and I was not disappointed this tour. The first one, the Imagery Mix Study, was a big effort but did not have the necessary ingredients for a confrontation between the Air Force and CIA. The second study effort however did lead to a showdown and again I wound up on the wrong side, this time siding with the CIA. The issue was one of modifying the NRT imagery satellite to expand its military operational support capability to one with a robust area search coverage capability so as to completely replace an existing satellite program–Hexagon. The two CIA reps on the study, Jimmy Ames and Marv Martin from

OD&E, were probably the two most intense guys I have ever worked with, very forceful and convincing about the benefits of forging on with improvements to the NRT imagery satellite system. The only problem however was that the system hadn't even flown yet, something which was pointed out to me by Mr. Howard and based on his more prudent appreciation of the issue, the Steering Group rejected the CIA idea. One of benefits of this effort was the beginning of a continuing relationship between myself and Marvelous Marv, as he was known in the community. He would leave CIA for TRW and eventually form his own company, and over the next ten years, we would get together off and on. One memorable occasion was dinner with our wives at El Bodegon on R Street, where for the first time I enjoyed drinking wine poured from a goatskin onto my forehead, a truly novel experience. Marv was the proximate cause of my 1982 hiring by E-Systems; close friends with Jack Maxey who ran its Center for Advanced Planning and Analysis, he convinced Jack to employ me as a Senior Member of the Technical Staff even though I was an 'English major.'

Defense Dissemination Program

SecDef quickly approved the NRT imagery dissemination scheme but who would develop the capability was not immediately decided. There were three contestants: DIA who dreamed up the idea but did not have an engineering staff to handle the job; CIA who owned the source of the imagery and could easily integrate the Washington end of the dissemination system into their infrastructure; and the Air Force, who were deep into in-theater imagery transmission and were not about to let the CIA run anything that had as its basis, a military requirement. I was still nominally in charge of the project and tried to influence it to some degree with help from two contractors, RCA and Philco Ford, both of whom had NASA imagery transmission contracts. Later that year the Program Decision Memorandum was approved giving the job to the Air Force Systems Command. I was assigned as an advisor to the Program Office and served in that capacity until late 1975. COL Ollie Fix, the Program Manager, was a real firecracker and didn't like to take guidance from anyone so the Air Force assigned naval academy graduate LtCol Don Cromer (later a LtGen who headed the Space Systems Division) to keep an eye on Ollie. Despite some rough times, the program office got the job done. Years later as a DIA civilian, I attended a staff meeting at the Military Airlift Command at Scott AFB (an add-on CINC) and watched the intelligence briefer flash imagery of the Soviet Union in front of the CinC that had been collected from space just hours before.

Deputy Chief Imagery Division

On the first of June 1975, I moved into the position of Deputy Chief, Imagery Division (DC-5) and went to work for Roy Sexton, a civilian who I had known from my previous tour in DIA. I was filling a Captain's billet and there was some expectation that I would soon be promoted to fill it. Roy was immediately pulled off for a special interagency assignment and right away I was running the division. It was a strange experience. Weeks before, I was at the bottom of my chain of command, and now suddenly, I was near the top. But my past lack of concern about titles and reporting seniors helped immeasurably in blunting any resentment that my former reporting seniors might have felt at my quick advancement. I ran the Imagery Division with the same enthusiasm I demonstrated in every other job I had. Four months, later Roy came back to work and shortly thereafter I would again be moved. In retrospect, I was just a blip in the imagery division leadership saga that went on for years to come but it was an enjoyable interlude that I will always cherish.

Promotion to O-6

In July 1975, I got a call at home from Admiral Harvey, the Director of Naval Intelligence congratulating me on being selected for Captain. I was aware that the Board was meeting but actually being on the list, was like all my other promotions, still a big surprise. As soon as the paperwork was completed in early September, I was frocked a Captain (entitles one to wear the uniform but not get the pay) which was the appropriate rank for the Deputy Imagery Division position. I had always been skeptical about promotion as evidence of doing a good job as I had come across many an O-6 who was worthless. But the idea that one got that way after being promoted, was a possibility that somewhat worried me. While I had enough confidence in myself to stay on an even keel, this last promotion was very satisfying. If nobody else knew anything specific about my work except for those directly involved, at least the Navy saw fit to publicly reward me. The photo on the right shows my wife and two youngest children, Frank Jr. and Alex, at my promotion ceremony. This picture, thanks to a DIA photographer, captures a very important moment in my life. I spent a little over two years in grade

before I retired and I would later have the opportunity to reflect on being a Captain compared to holding an equivalent GS-15 civilian rank. Pretty much the same pay and responsibilities, but having the uniform on was a bonus; somehow or other, it made a big difference in how I went about getting things done.

Trip to Europe

In October 1975, I traveled to Europe with Mr. Hughes to visit the Supreme Allied Commander Europe (SACEUR) and other senior U.S Commanders in the theater. John liked having a navy Captain as his bag holder and he continued to value my counsel on reconnaissance matters. The ostensible purpose of the trip was to give the same briefing on the Soviet threat that he had successfully provided Congress but it was not lost on any of my shipmates that Oktoberfest was in full swing. We took a commercial 747 direct to Brussels, made interesting from a security standpoint by Mr. Hughes in-flight review of his briefing slides, all classified Top Secret Codeword. When we got to SACEUR headquarters at Mons just outside Brussels, we were disappointed that the man himself, General Alexander Haig, former White House Chief of Staff and future Secretary of State under President Reagan, was out in the field with the troops. But we sat down with the other members of the staff who were appropriately cleared, and presented the briefing.

Then over the next week, we traveled around Germany to Stuttgart to brief the Commander in Chief of US Forces (CINCEUR), then to Wiesbaden to do the same for the Air Force biggies, and finally Heidelberg for the top Army brass. At one time or another, most everyone of importance on these staffs had either worked with or in some way been associated with John, and we were treated royally. The trip to Heidelberg was especially enjoyable as we got a grade A tour of the city with all its historic landmarks. I had been to southern Europe in the mid 1960s traveling on my own and what an improvement it was being with a VIP. One other nice thing that happened was a job offer to be the Deputy J-2 on the CINCEUR staff. The only problem was that I had to take the job immediately and I just couldn't figure out how to do that, given all I was involved in back at DIA. Just as importantly, I did not think the assignment would be good for the family, especially in respect to the stability of the kids schooling. So I politely said no and as you will read below, in a sense lived to regret it.

After we had finished up briefing the senior US Commanders, we hitched a ride in an Army puddle-jumper to catch up with GEN Haig in Kitzingen. A Joint U.S. and German field exercise was ending that day and we had every expectation of catching a ride back with the General (and his wife) to Belgium. And indeed we did. I had the pleasure of sitting in GEN Haig's compartment and listen to him and John candidly discuss a number of defense issues; one that stuck in my mind was Haig's preference for the Germans as a fighting force as contrasted with other NATO units including the US The photo on the right shows me exiting SACUER's plane when we got back to Belgium. The only bad part of the trip was the flight home the next day. The Lufthansa 747 out of Frankfurt couldn't get off the ground and rather than wait around another day, we made some adjustments and got seats on a Sabena 707. The flight was hours longer and the seats so small and cramped, it was almost torture. And then we had a problem getting back to DC from New York. So there can be a downside to some otherwise very enjoyable VIP junkets.

Directorate for Collection Operations

In March 1976, I was appointed Chief, Current Operations Division (DC-1) and the point of contact with the JCS/JRC for all intelligence collection requirements and reconnaissance contingency plans. And guess where my new office was located? The same spot that Col Humphrey's sat in 10 years previously. But right away there was a problem, since a senior Air Force Colonel had historically filled the slot, and now it was being given to a junior Navy Captain. I would never have known about all this if not for the close relationship with the JRC Air Force guys next door who were aware of the situation and kept me abreast of the political infighting at the higher levels. But Mr. Hughes was adamant that he wanted me in the slot and he now had enough horsepower to pull it off. When I left DIA later that year, it reverted back to an Air Force Colonel's position. In any case, the job had three major responsibilities:

Oversee three briefing officers who kept senior DoD and Joint Staff officials up to date on current collection efforts. Each morning they would get in early, develop the charts, photos, etc., and then set off to participate in the morning command briefings in the Pentagon. I tried somewhat unsuccessfully to expand their expertise to address weather problems, especially as they factored into imagery collection missions, but never got too far in that. I also tried to get them to better understand the relationship of the various collection disciplines so as to more intelligently answer questions that invariably came up at the briefing, and in that I failed miserably. I am not sure whether this was a motivational problem or the lack of some inherent characteristic, i.e., not be able to grasp nuts and bolts matters like, in the case of satellite collection, orbital mechanics.

Develop and staff the J-2 JCS position on the Peacetime Reconnaissance and Certain Sensitive Operations Program. Each month, the services and defense agencies were required to submit their plans to the NSC for reconnaissance missions that might cause an incident. The program was initiated right after the capture of the Pueblo by the North Koreans, where the post mortem showed top government officials were unaware of what the Pueblo was doing and why. The J-2 part of this monthly drill was to certify that the gains outweighed the risks, and I don't think I have ever done as much soul searching as when I had to sign my name saying the reconnaissance schedule for the month looked good and recommend the Director DIA, acting as the J-2, sign off on it. I focused a lot on navy submarine missions where the intelligence product was very compartmentalized and the risks great. I would recommend anyone interested in this subject read Sherry Sontag's *Blind Man's Bluff*. I never had anything backfire but I would wake up many a night and wonder whether or not I had done the right thing.

Get the Collection Coordination Facility up to speed assuming its mission in the new near real time collection environment. The facility was located directly below my office in the National Military Intelligence Center and staffed by junior grade officers with some experience in one or another of the collection disciplines but no exposure to anything resembling an all-source environment. After finishing up my normal work day, I would mosey down to the CCF, observe, instruct, and in despair, go home at night having missed supper for the hundredth time and moan about how the most brilliant concept can come unglued if you do not have the right people to execute it. But I hoped that with a sound concept, eventually the need for competent staffing would become apparent and things would improve.

The standard routine part of the job involved coordination on miscellaneous JCS/DIA collection matters. One such incident had me sitting down at my typewriter and completely rewriting a letter from CJCS to SecDef, as the action officer, an Air Force LCOL from another division sat in amazement. Years later I met him at a social function and he remembered the incident saying he had never seen anything like that kind of interest being shown by anybody on any of the papers he trotted around trying to get approved.

Viktor Belenko/MIG 25

In my final DIA fitness report, VADM Inman wrote about my role in the "coordination, tasking and monitoring of a sensitive exploitation arrangement which required daily reporting to the SecDef and the Director, DIA" This all came about when Soviet pilot Viktor Belenko landed a MIG-25 Foxbat intact at Hakodate Airport in Japan (photo at right) and asked for asylum in the United States. At that time the MIG-25 was considered the most advanced interceptor in the world, and when it set down in Japan, a feeding frenzy began to take place, as US technical experts scrambled to get access to the pilot (CIA had the handle on this), the pilot's manual, and the aircraft itself to ascertain exactly how formidable a plane it was. The photo below shows a Foxbat in flight.

Resting just off the runway in the grass, on site examination of the airframe, power plant, communications, radar and weapons systems began immediately. During the next two months, the Russians put every bit of pressure they could to get it returned to them (eventually it was in crates) as the US tried to get more and more technical assessment teams out to the aircraft, while the Japanese in a diplomatic bind, temporized with the Russians at the same time providing the US full access to the aircraft. Recognizing early on that inter-service, inter-agency rivalry could impede the proper examination of the aircraft, Inman as the Deputy Director of DIA assigned me to head a task force charged with coordinating the exploitation effort. Herding cats is the best way to describe my efforts with this group. The only thing I did well was report up the line on what was happening. Every morning at 0500, I began to recap what the teams in Japan had done, read the message traffic on the Soviet-Japanese diplomatic negotiations, tried to guess how many more days we would have access to the aircraft and exactly what part of the exploitation should take priority in this time period. I put all this in a letter to Secretary of Defense Donald Rumsfeld (yes the same one!) and hand delivered it to one of his executive assistants, Col. Elmer Brooks at 0700. Composed on the fly as I absorbed the reporting, I typed the letters myself and signed them for the Director DIA.

There was one problem however; I was so busy with the whole enterprise, I forgot to tell either Mr. Hughes or the DIA Director what I was doing. I can remember the shock on his face when Mr. Hughes first found out. Standing in the doorway of my office and looking at me quizzically as I told him I had been writing letters every day to the SecDef in the name of the Director and somehow I had forgotten to tell him. "You've been doing what?" he said amazed. But like most of my shortcomings, it was one of commission rather than omission and someway or other, most of my bosses learned to live with that. The MIG-25 exploitation effort turned out to be a great success by the way. The overall intelligence gained gave the US a much clearer picture of Soviet technology, something I characterized at the time as 'crude simplicity.' And Viktor Belenko melted into American society but not before helping John Barron write *MIG Pilot* about his experiences.

Moving On

In late October 1976, VADM Inman called me up to his office and told me I was being ordered to the Intelligence Community (IC) Staff to become a Special Assistant to the Director of Central Intelligence who at that time was George H.W. Bush, inferring I had been asked for by name. The incumbent, CAPT Ron North had already departed and my orders were being written as we spoke. I was not all that happy about this, as I felt that getting the CCF guys up to speed on doing their thing was more important. But I said "Aye, Aye" with a smile and went back to my office to clear my desk in as professional manner as I could. Thinking back about the matter with the Air Force wanting the billet filled with one of their own, and my less than politic handling of the

MIG-25 exploitation effort, it dawned on me that maybe I was fired. One of my close associates, the new Deputy Director for Imagery Gene Bacher, thought that while this might be the case, it looked more like a promotion in the sense that I was heading into the world of big time intelligence shakers and movers. In any case, I was so wrapped up in doing things, I didn't spend any time worrying about it.

I should note however that four years later, when I was working in DIA as a civilian, ADM Inman, then Deputy Director of CIA, did effectively remove me from an Interagency Task Force he had commissioned to look at Intelligence Capabilities, 1985-1990. While I got reinstated on that project primarily because DIA had no one else who could actually do the job, Inman clearly had mixed feelings about my performance. But when I left DIA in late 1976, Mr. Hughes pushed through the paperwork for another Meritorious Service Medal and in almost all respects, my second tour in DIA was a great success.

Chapter 9-Intelligence Community Staff (1976-1978)

Working for the CIA?

Even though I was assigned to the Director of Central Intelligence (DCI) staff, I was treated like a new employee of the CIA for administrative purposes and required to take a polygraph as well as a battery of psychological tests. I very much enjoyed the polygraph experience, especially talking to the examiner about the way he structured questions and how I figured out the best way to answer them. For example, the question "Have you ever lied to your superior?" came half way in the exam and by that time I began to understand how to respond. In this instance, I laughed saying "Yes,-doesn't everyone." He then asked me to elaborate, and after going back and forth for a while, we left the subject agreeing that if one wanted to get the job done, sometimes you had to either keep your boss in the dark or if necessary, lie to him. For the psychological examination, I made a point of wearing my uniform to the testing site, where about twenty of us took a written exam. I made no attempt to carefully answer the questions, interested only in completing the test as quickly as possible. When I finished and presented it to the proctor, a young woman psychologist, she did a quick scan, looked up at me with a smile and said "You already have the job, don't you?"

Another quirk of the assignment was a set of military orders that permitted me to go anywhere in the world, for any reason at any time, and have it all paid for by the government (see Appendix B). While I never used them, it made me realize I was now operating in a different world than before. The only down side to the new assignment, at least initially, was the extended drive to Langley, Virginia where the IC Staff was headquartered. After winding my way up the Potomac from Tantallon to the Wilson Bridge, I had three route choices. The shortest was to cross the bridge and go north through Old Town Alexandria and past the Pentagon to the GW Parkway. The second shortest took me up I-295 into D.C. and across the Memorial Bridge and then to the GW Parkway. The third and longest way was to head west and then north on the beltway to the southbound GW Parkway exit or come back east on local roads to CIA Headquarters. I routinely tried all three ways in an attempt to avoid traffic and for a time took the bus (my VW Bug had some problems) but never succeeded in making the commute anything other than hell.

Welcome Aboard

Effective early November, my assignment on the IC Staff was to the Office of Policy and Planning (OPP). Our spaces were located on the sixth floor with windows facing the front of the building, with quick access to a stairwell just down the corridor from the DCI's office on the 7th floor. I initially had a rough couple of months with my reporting senior, BGEN Ted Atkeson, who was not sure what he had in me. That previous summer I had rejected an informal overture to join the staff; the recruiters Howie Shue and Paul Janota had accurately related back to him my reason "Why in God's name" I had said "would anyone want to work for a bastard outfit like yours?" I would find out later he shared my views but I am sure at the time he wondered whether he had a problem case on his hands. But watching Admiral Dan Murphy (Director IC Staff) pin a Meritorious Service Medal on my chest for my DIA work and

meeting Mary at the awards ceremony helped very much in warming up our relationship. In retrospect, it was a situation in reverse to visiting Ned Humphrey's house 10 years before, meeting his spouse, and then recognizing that somebody with such a nice wife could not be all that bad. Ted also hosted a newcomer party where I met many old CIA hands as well as the wives of my new shipmates. I related well with all of them; only two other OPP staffers were active duty military but the rest were either ex-military or CIA, all smart, and one, Bob Sorley, mentioned in the introduction, would soon write a marvelous book on Vietnam,-*A Better War*. So all in all, it was a good start to a job that I was not all that keen about getting.

Building Blocks

Fortunately, the work environment got intense very fast and my concerns about the job disappeared. The enthusiasm and assuredness I directed toward what I knew best, the National/Tactical Interface, quickly established my professional bona fides and got me out of the rookie category. In early 1977, Congress directed the DCI to form an interagency working group to look at the interface problem, an effort I co-chaired with Wally Henderson, an OSD civilian. We started immediately and in one way or another, it occupied all my time for pretty much that entire first year on the staff.

Before I get to that, a word about my bosses. General Atkeson, once I convinced him I wasn't a loose cannon, supported all my efforts in whatever way he could. At one time, he directed all of the OPP resources to directly support my efforts on the Nat/Tac effort. He also helped me in my writing, basically putting all that I had learned in school about outlining into a usable perspective. It went like this: first think about the three most important things you want to say about a subject and in a phrase, jot them down. Then take each of the three points and in a phrase, list the three most important things you want to say about each of them and jot them down. Then at this point, and only then, should you start to write, by trying to logically tie the nine phrases together. What was surprising was here I was in my mid-forties learning for the first time how to more effectively express myself in writing. To give you an idea of his writing ability as well as some insight as to how he looked at my work, his direct, forceful and parsimonious verbal assessment of my performance in my Navy Fitness report read as follows:

> Captain Bausch is an exceptionally effective officer who knows his business cold. In my estimation, he has a deeper understanding and better grasp of the complexities of the national/tactical intelligence interface than anyone else on the Intelligence Community Staff. He speaks with authority, he makes solid good sense and he presses his views and analyses with vigor. He is tough, intelligent and imaginative. He would make an outstanding flag officer and should be selected for promotion to Rear Admiral now.

As for my big bosses, the few times I got to interact with Mr. Bush, I was impressed. Courteous and thoughtful, he was a wonderful person to work for. ADM Murphy, the head of the IC Staff was a brand new four-star, a former Sixth Fleet Commander and would go on to be Chief of Staff to Mr. Bush when Bush became Vice President. His career CIA Deputy, John McMahon, was the former head of the Office of ELINT (OEL) and the Office of Technical Services (OTS), and future CIA Director of Operations, Director of Intelligence and CIA Deputy Director, and eventually President, Lockheed Missiles & Space. Both Admiral Murphy and John were professionals to the core and easy to work with. But things changed quickly.

When a Democrat, President Jimmy Carter, took over the White House in January 1977, Mr. Bush, a Republican was shoved out the door. Then after a period of uncertainty in naming his replacement, (the CIA Deputy at the time, Hank Knoche, acted as DCI) ADM Stansfield Turner was appointed to the position. In May 1977, ADM Murphy left to be Deputy Under Secretary Defense for Policy (having two full Admirals at the top of the intelligence hierarchy was considered overkill, or more likely, the ruling powers envisioned fierce infighting for power between the two). But John McMahon stayed on as Acting Director of the IC staff and was instrumental in keeping the hounds at bay as I ground out the Nat/Tac study. In March 1977, the IC Staff moved out of CIA to 1724 F St. Wash DC which prompted the humorous Bill Safire article on the right. This move was prompted by Congress, in this particular instance, evidencing concern about maintaining the independence of the IC Staff from the CIA.

The House on F Street

By William Safire

Memo to: Comrade-President Brezhnev
From: K.G.B. Washington Desk
Re: Analysis of proposed reorganization of American intelligence community by President Carter.

Penetration has not yet been effectuated of "The House on F Street," former Selective Service System headquarters two blocks west of the White House now taken over by an elite corps from C.I.A. seeking to take over entire U.S. intelligence community. But certain assumptions can be made:

1. President Carter, like all his predecessors, is determined to have best possible intelligence-gathering and evaluations to help him avoid foreign policy blunders.

2. To obtain that, he desires to centralize operations, budgeting and planning authority under one "czar" (pardon expression) whom he personally trusts.

3. Such a centralized organization would be designed to reduce internal intelligence arguments and diminish the number of people with access to strategic information, thus reducing danger of leaks.

4. To reassure people that this tightening-up will not lead to "abuses," he will make much propaganda about uncovering past C.I.A. mistakes and laying down guidelines for protection of rights of persons cooperating with K.G.B. Washington.

In answer to your specific questions, Comrade President:

1. *Why will Carter's generals permit him to remove the capacity for military intelligence from Defense Department?*

Although removal of budgetary and planning control of Defense Intelligence Agency and National Security Agency from the Defense Department has been opposed publicly by Defense Secretary Brown, expected fury has been moderated by:

(a) Old-boy military network. Proposal of Carter Annapolis classmate Admiral Stansfield Turner to be boss of all intelligence bosses cools down Pentagon, which expects him to be named chairman of Joint Chiefs of Staff in couple of years (pointedly, he did not resign naval commission).

(b) Anti-old-boy network. Officials at N.S.A. (who had been irritated with Ford White House counsel Philip Buchen for cutting extent of electronic surveillance of resident aliens), as well as other non-C.I.A. intelligence units, want the Director of Central Intelligence removed from basic job as head of C.I.A., on the theory that without the long-term bureaucratic backup of a large agency, the D.C.I.'s power can be nibbled away.

2. *What evidence exists for our centralization assumption?*

(a) Pattern of previous behavior. In reorganizing the energy program, Mr. Carter chose to centralize power in one man, the hawk Schlesinger. Disrespectful wags here say "Brezhnev likes the title of President, Carter likes the title of czar."

(b) Abolition of "piffiab." The President's Foreign Intelligence Advisory Board was the only group of citizens capable of questioning intelligence estimates and taking challenges direct to President. It once even arranged for an academic group's alternative to the official strategic estimate, embarrassing some at C.I.A. On Admiral Turners advice, PFIAB was recently abolished, drawing tighter the circle of access to the President.

Reason why little editorial objection was raised is a confusion of PFIAB's function (of improving evaluation) with that of the new Intelligence Oversight Board (of checking on abuses). Most people think PFIAB failed to limit abuses, which it was never in business to do, and that the smaller I.O.B. will now question intelligence analysis, which it is not in business to do. Analysts who disagree with D.C.I. "czar" will soon have no way to get to President.

3. *Why will Americans, after the Watergate and C.I.A. investigations, hold still for the concentration of intelligence power?*

(a) Americans today are more concerned about the C.I.A.'s *doing* wrong than the C.I.A.'s *being* wrong. Like generals fighting the past war, senators limit their oversight to methods rather than conclusions, to operatives rather than product.

(b) Mr. Carter is exploiting this preoccupation with the past. His Justice Department indicted an F.B.I. official for too-zealous intelligence work (knowing the man is bound to be acquitted). And the inclination not to prosecute the C.I.A cases has been replaced by a new zeal to nail Richard Helms: Prosecution of past "abuses" is a perfect cover for promotion of present centralization, making tightened security procedures and a more isolated President seem to be part of "reform."

All of us here in K.G.B. Washington —from the station chief driving Ambassador Dobrynin's car to the agents in place across the street from the House on F Street—send you heartiest congratulations on your assumption of the Soviet Presidency, along with the suggestion that you use your new authority to wrest control of military intelligence from the Red Army bureaucracy and concentrate it under the personally loyal chief of the K.G.B., who always has your best interests at heart.

The New York Times
Published: June 23, 1977

The 1977 National/Tactical Interface Study

My experience prior to getting involved with the Nat/Tac study effort fortunately gave me firm footing that sustained me from the start. Having worked the problem in the past from both ends, I believed that there was no organizational solution, something Congress very much wanted. In general, the Congress defined the issue as a bureaucratic one and not for the last time, the idea of a Director of National Intelligence surfaced. Seeing the problem differently, I worked hard at structuring an understanding of the issue that would help me prove my case. First I argued that the flow of tactical intelligence upwards to the national level was not a real problem; each and every example of a deficiency could be remedied by command attention, more money and most of all, freedom to act on solving it by dedicated competent, professionals. This latter approach by the way is my general solution to almost every problem of this kind. The flow of national intelligence downwards to the field commander was indeed a problem, primarily because of security concerns, insufficient communications bandwidth between higher headquarters and the field, and inadequate processing and display equipment in theater to meet tactical commander's requirements.

The overall study effort had two phases: first, a report by the Secretary of Defense/Director of Central Intelligence to Congress on the subject, to be followed by a Defense Science Board (DSB) investigation of the problem and critique of the SecDef/DCI report. As for the first report, it did a thorough job of explaining both sides of the problem, laid out the pros and cons of potential fixes, and touted things like the Defense Dissemination System. In general, the report recommended expanding dissemination of NRP imagery and SIGINT to tactical users, a process which if prudently and aggressively pursued, would over time, eventually satisfy those tactical intelligence needs for national reconnaissance products. I don't think anybody was especially happy with the bottom line but it was a professional effort and no one to my knowledge disputed it findings. Upon its completion, a DSB panel was formed with top-notch membership. Though the Chairman, John Baldeschwieler, a chemistry professor from CalTech was a rookie, he was shepherded by Gene Fubini an old hand at these type efforts and as a result, John did fine in his leadership role. The other members included a broad section of the nations top civilian scientists involved in the intelligence and reconnaissance business, one of whom was Dr. Lloyd Lauderdale. Lloyd was a former CIA Deputy Director for Science and Technology, the driving force behind RHYOLITE, had worked with John McMahon in the late 1960s and currently was the Vice President of Research for E-Systems. Five years later, he would okay my hire to work in its Center for Advanced Planning and Analysis.

The DSB panel believed in getting its facts first hand from the tactical users, and accompanied by DoD's Wally Henderson and myself, we toured the country talking to Army, Navy, Air Force and Marine commanders. One visit to Fort Bragg NC (18th Airborne Corps) was an eye-opener to some of the more sheltered scientists. Invited to the Officer's Club, they witnessed juiced up paratroopers jumping off tables yelling 'Geronimo,' then rolling along the floor and rising erect at perfect attention. The DSB effort was extended into 1978 to address an added twist, the National Intelligence Tasking Center (see below), and eventually another report was produced which pretty much recommended staying the current course. From my perspective, it was nice to have my original report findings confirmed and of course I loved hobnobbing with these bright, very accessible, science oriented guys.

New Faces

Admiral Turner, hoping to change and improve the intelligence community and gain more control over it, relied on, to quote a critic, "his own intellect and self-confidence and on aides who themselves did not have intelligence backgrounds." When Admiral Turner finally started to put his people into the key IC Staff positions, he first replaced John McMahon, who was acting as head of the IC Staff. He chose a brilliant civilian from the Congressional Budget Officer, John Koehler. John was a Rand Corporation alumnus, had a sharp mind, very savvy on budgets, comfortable with any kind of data analysis and emphasized "outcomes" when looking at problem solving. I liked the way he saw the world but on the down side, he knew nothing about military operations or about reconnaissance. And in a short time, what little we might have found in common was soon sundered with the bifurcation of the IC Staff. It also did not help that the first real challenge the resource management part of the IC Staff was forced to address under John's aegis was zero based budgeting (ZBB), a method where all expenditures had to be justified each and every year, as if they were all new programs. President Carter had become fond of it when Governor of Georgia and in 1977 ordered its government wide implementation.

The National Foreign Intelligence Program (NFIP) whose budget we were responsible for got special attention and to his credit, John played the game straight and followed the guidelines set forth in OMB Circular A-11, unlike many other branches of government. But the increase in seemingly unnecessary documentation required to meet the goals of the budgeting system were extensive and one of the very first things the Reagan Administration did in 1981 was to kill ZBB. The only really controversial FY-1978 NFIP budget issue was a major space related one, again between CIA and the Air Force concerning a new reconnaissance system but that became so mired in the IC Staff reorganization politics that OPP office never even participated in the debate.

Splitting the IC Staff–the NITC

The embedding of a Collection Coordination Facility in the National Military Command Center somehow caught Admiral Turner's attention. Broadening and expanding the concept, he convinced President Carter to first issue a Presidential Directive in August 1977 followed up in January 1978 by Executive Order 12036 establishing the National Intelligence Tasking Center (NITC). What was surprising about all of this was that there were no studies or analyses performed on the merits of an NITC and no provision for quarters or staffing. It was just an idea and it was up to Admiral Turner and the intelligence/defense community to figure out how to execute it. And then to compound the problem, Admiral Turner brought on board LGEN Frank Camm, a retired US Army engineer, to flesh out the NITC idea and to head that part of the IC staff involved with intelligence collection: COMIREX, the Signals Intelligence (SIGINT) and the Human Intelligence (HUMINT) committees. I had a golfing buddy, a retired BGEN tanker, who worked with Camm and the best he could say about him was that he was an 'idiot.' I never worked close enough to him to validate that representation but his efforts to develop the NITC were ineffectual. He became confused easily, his portion of the IC Staff were not structured to help in his bureaucratic adventure and every once in a while, he would broach a wild idea like turning over tasking of national reconnaissance satellite systems to tactical commanders during combat operations. And along with all these problems, he got into a pissing contest with Joint Chiefs of Staff about his need for a Pentagon office on the E-Ring with a window, something he put as his top priority (but never got). Like ZBB, the NITC never got off the ground and the concept was abandoned when President Reagan took office. To postscript this story, four years later as a DIA civilian, I would sit in my office on the E-ring looking out my window and smile thinking about General Camm and his NITC travail.

Space Shuttle

Back in the late 1960s, my DIA boss Mr. Howard, trying to broaden my understanding of the space business, sent me off to a NASA sponsored colloquium on Manned Space Flight. The only speaker I can remember from the event was Amron Katz, who made a passionate case against the very subject the meeting was addressing. His argument was that manned space flight was not safe and could never be made so, and that the costs of attempting to do it would be so large as to eventually hurt the space program. This subject surfaced in the national security community in 1977, as it wrestled with what to do about expendable launch vehicles (ELV), which the Air Force used to put all classified payloads into space. NASA, needing DOD's business, urged the White House mandate all these payloads be launched on the Shuttle rather than ELVs. I remember sitting in on sessions when this issue was addressed where experienced space warriors like NRO Staff Director Jimmy Hill would point out the likelihood of a shuttle disaster and the inevitable launch hiatus for vital national security payloads. Despite their best efforts to convince him otherwise, Admiral Turner, then deeply involved in his jurisdictional clashes, did not fight NASA nor intercede with President Carter on this issue. There is no mention in Admiral Turner's book, *Secrecy and Democracy*, of this matter at all, and I find that particularly instructive. He was so concerned about changing things, he completely forgot about what worked and what didn't need a new approach. Despite all the arguing that the shuttle would never prove that reliable, an executive order came out of the White House which effectively required NRP assets be launched on the shuttle.

The photo above shows an early 1980s shuttle 'fit test' at Space Launch Complex 6, Vandenberg AFB CA. Because of payload weight and launch inclination angles, substantial changes to many launch pads were required; SLC-6 underwent years of reconstruction and never launched a shuttle (modification costs have been estimated at $3 billion). The Air Force then throttled back on ELV production and NASA had to make changes in command and control security at the Johnson Space Center (JSC) in Houston to handle the launch of highly classified payloads. Years later after national reconnaissance spacecraft were configured for the shuttle, as Amron Katz might have predicted, the Challenger disaster occurred. Faced with the unacceptable delays needed to get the shuttle orbiters back in space again, the Reagan administration decided to go back to ELV's for vital national security payloads. By that time, the cost for this shuttle misadventure was in excess of $20 billion. And as one Defense Department critic noted at the time for those that might be mentally challenged: "mission accomplishment was adversely affected."

As for the main protagonists, Admiral Turner seems to have avoided any public criticism (I suspect some of his other shortcomings were more targetable) and as for Jimmy Hill, Dwayne Day in the Space Review (www.thespacereview.com/article/748/2) wrote: "people knowledgeable about these issues have confirmed that Hill fought the shift of NRO spacecraft to the shuttle, and ...part of the reason that Hill has earned almost cult-like status within the agency is because the Challenger accident later proved him right."

Security at the Johnson Space Center

Following the 1986 Challenger launch disaster, the Columbia's 2003 disintegration in space settled for all time who was right in the shuttle versus ELV contest. But it was not the only one of Admiral Turner's shuttle related missteps. In mid 1978, he appointed me to be a part of a Aeronautics and Astronautics Advisory Board effort to define and then cost the aforementioned necessary security changes needed at NASA's Johnson Space Center (JSC) in Houston. I represented the Resource Management part of the IC Staff while Roland Inlow, still Chairman of the COMIREX, acted for the Collection Tasking side. From an interest standpoint, it was winner. Interfacing with payload managers, the Aerospace Corporation and NASA was an extremely rewarding experience. One particular episode sticks out in my memory. After working the day in my F Street office, I boarded a late afternoon flight to Los Angeles to attend an early evening (west coast time) conference to review our progress. Then I got back to the airport in time to take the late evening redeye to return to Washington so as to show up for work on time the next morning. I do not have a documented record of the costs to reconfigure the JSC but the figure of $120 million sticks in my mind. And if that is correct, it turned out to be relatively cheap—less than one percent of the Space Shuttle/ELV fiasco's total cost.

Ten Months in Purgatory

General Atkeson left the staff the last day in February 1978, happy to return to the Pentagon. And he did me what he thought was one last favor; based on his recommendation, I took over his job as the Chief, Policy and Plans, a position calling for a Flag Officer or equivalent. I immediately started work on a project I had been thinking about for a year, the National Intelligence Plan (NIP). Modeled after the well thought of Defense Intelligence Plan, the first draft of the NIP was poorly received and I quickly gave up on the idea. The rest of 1978 became an agony for me in respect to the core of my IC staff related functions. Lack of meaningful role for OPP in the NFIP budget process was one major disappointment. Zero based budget decisions were now being made in smoke filled back rooms to which, as a planner I was not invited. More importantly, serving witness to the machinations involving the NITC and Frank Camm became almost more than I could stand.

I did have one small consolation in that I attended DCI staff meetings when John Koehler was out of town. I was driven back and forth from F Street to CIA Headquarters by a professional CIA driver and bodyguard and I loved to hear him talk about the offensive and defensive driving skills he had learned at school, e.g., how to nudge a car off the road with just a gentle tap. Attending the staff meetings was less fun but an eye opener, inasmuch as I saw issues being discussed and strategies being formulated to include needed coordination between the DCI and senior DoD officials.

Through my Pentagon connections, I would read back-channel communications between and among these officials about these issues; what shocked me were the substantial differences between what Admiral Turner as opposed to the Pentagon Admirals and Generals understood on many of these matters. And I am sure I was not the only one at the staff meetings who were aware of this. Tension, when such matters were broached, was palpable. Watching Frank Carlucci, the Deputy DCI, who would later become Deputy Secretary of Defense, deal with Admiral Turner as contrasted with a long time CIA guy like Chief of Administration, Jack Blake was illuminating. Both appeared to humor Admiral Turner but with different 'this too shall pass' outlooks. And I of course mostly listened and wished I worked somewhere else.

Decision to retire

In September of that year, I pretty much decided to abandon ship. Before Ted Atkeson left to go back to the Pentagon, we had talked about retirement, specifically the financial aspects. Given that I could then retire on 60% of my base pay and even with the loss of allowances, he figured I was working for half pay. I had also done some quick math regarding the projected cost of college education (two of the children were already in college and three more on deck) and for the first time in my life, I started applying monetary considerations to what I thought about doing. I fretted a little about that, as well as many other things during my 5-10 mile marathon training runs at lunch-time. I was also thought about potential future assignment choices, e.g., appealing ones like Hawaii as Commanding Officer FICPAC followed by J-2 CINCPAC and not so appealing ones like a 13 month unaccompanied tour as the N-2, US

Naval Forces Korea. After seeking some further counsel from my contemporaries and getting advice from VADM Inman (then head of NSA), I put in my request for retirement. My active duty career ended on December 31, 1978 and the next day, I was transferred to the retired list.

Postscript

My relief as Chief of OPP did not come aboard until late March 1979 so I stayed on as a civilian under contract to CIA for the first three months of the year. It was then that I first read *The Peter Principle* and after some self-examination, recognized that I was one of those who had been promoted to a position where they are no longer effective, i.e., reached their 'level of incompetence.' But it was not something I felt too bad about; I was so happy to get out from under. The IC Staff experience was one I would never forget and gave me a new understanding of the Serenity Prayer. Now I saw it as a paean to people who have the wisdom to solve problems that do not exist and then crow about how well they did.

In April 1979, I hired on as a GS-15 civil servant with DIA and went to work as a Special Assistant to Mr. Hughes, then the new Deputy Director for Defense Intelligence. I toiled there until the end of 1981 when I became salary capped. I would have liked to have stayed but I needed money for college tuition so in early 1982 I got a job with E-Systems as a Senior Member of the Technical Staff at its Center for Analysis and Planning. This was made possible by my previous work in the NRP and after ten years with E-Systems, which qualified me for a pension and other benefits, I retired the second week in February 1992 thus ending my professional career.

Appendix A. Awards/Citations

THE SECRETARY OF THE NAVY
WASHINGTON

The Secretary of the Navy takes pleasure in commending the

UNITED STATES NAVAL PHOTOGRAPHIC INTERPRETATION CENTER

for service as set forth in the following

CITATION:

"For exceptionally meritorious service from 1 May through 31 December 1962. During the period of international crisis, the United States Naval Photographic Interpretation Center received, reproduced, and conducted immediate analysis and timely distribution of intelligence photography to the President of the United States, his principal policy advisors, and members of the U.S. Intelligence Community. The Center, by rapid adjustment to the emergency, and through diligent and skillful application of professional techniques, assisted to a notable degree in the identification of the nature and magnitude of the threat to world peace. Although working under unusual demands and great stress, personnel of the Center were able to analyze and process a massive volume of critical intelligence material, thereby contributing to the immediate and effective response by the United States to the developing threat. The outstanding achievement of the Naval Photographic Interpretation Center reflects great credit upon each member of the organization and upon the United States Naval Service."

All personnel attached to and serving with the United States Naval Photographic Interpretation Center during the above period, or any part thereof, are hereby authorized to wear the Navy Unit Commendation Ribbon.

Secretary of the Navy

DEPARTMENT OF DEFENSE

THIS IS TO CERTIFY THAT THE SECRETARY OF DEFENSE
HAS AUTHORIZED THE AWARD OF THE

JOINT SERVICE COMMENDATION MEDAL

TO LIEUTENANT COMMANDER FRANCIS A. BAUSCH, UNITED STATES NAVY, 593704/1350

FOR outstanding performance of duty while serving as Photographic Intelligence Officer on the Staff of the Commander in Chief Atlantic from 12 July 1963 to 11 October 1965. Displaying outstanding initiative and resourcefulness, Lieutenant Commander BAUSCH organized, planned and supervised the photographic efforts of the Atlantic Command. Through his comprehensive knowledge of intelligence collection management problems, he was very instrumental in establishing the requirements for the Atlantic Command Intelligence Collection Management System for the Automatic Data Processing Program. Throughout his tenure of duty he continually demonstrated outstanding leadership, professional competence, and devotion to duty thereby contributing significantly to the successful accomplishment of the mission of the Atlantic Command. Lieutenant Commander BAUSCH's outstanding performance of duty reflects great credit upon himself, the United States Navy, and the Atlantic Command.

GIVEN UNDER MY HAND THIS TWENTY-FIFTH DAY OF JANUARY 1966

T.H. Moorer

T. H. MOORER
Admiral, U. S. Navy
Commander in Chief Atlantic
for SECRETARY OF DEFENSE

The Atlantic Command
COMMAND OR OFFICE

THE SECRETARY OF THE NAVY
WASHINGTON

The President of the United States takes pleasure in presenting the LEGION OF MERIT to

COMMANDER FRANCIS A. BAUSCH
UNITED STATES NAVY

for service as set forth in the following

CITATION:

For exceptionally meritorious service from May 1968 to June 1971 as a Staff Intelligence Officer, Defense Intelligence Agency (DIA). Commander Bausch gave unstintingly of his time and energies in managing a most sensitive and demanding program which provides the highest authorities with products for decision-making. Based on the results of his efforts, national command authorities were better able to make decisions affecting national security and foreign policy. Commander Bausch was instrumental in formulating policy guidance for the national intelligence committee on which he served as the DIA member. Within DIA, he developed the idea of collection area specialists to provide a source of information for detailed intelligence needs. By his outstanding leadership, judgment, and inspiring devotion to duty throughout, Commander Bausch upheld the highest traditions of the United States Naval Service.

For the President,

John H. Chafee

Secretary of the Navy

CITATION TO ACCOMPANY THE AWARD OF

THE MERITORIOUS SERVICE MEDAL

TO

FRANCIS A. BAUSCH

Commander Francis A. Bausch, United States Navy, distinguished himself in the performance of outstanding service to the United States as Commanding Officer of the Naval Unit and as Deputy Director of the Armed Forces Air Intelligence Training Center, United States Air Force School of Applied Aerospace Sciences, Lowry Air Force Base, Colorado, from 19 June 1971 to 29 June 1973. During this period, Commander Bausch's professional knowledge, leadership, initiative, and resourcefulness have been directly responsible for the strengthened role of the Naval Unit at Lowry Air Force Base, the greatly enhanced scholastic reputation of the Armed Forces Air Intelligence Training Center, and the increased professional quality of air intelligence training for the Department of Defense. The singularly distinctive accomplishments of Commander Bausch reflect great credit upon himself, the United States Navy and the United States Air Force.

THE SECRETARY OF THE NAVY
WASHINGTON, D. C. 20350

The President of the United States takes pleasure in presenting the MERITORIOUS SERVICE MEDAL (Gold Star in lieu of the Second Award) to

CAPTAIN FRANCIS A. BAUSCH
UNITED STATES NAVY

for service as set forth in the following

CITATION:

For outstanding meritorious service while assigned to the Defense Intelligence Agency from April 1974 to November 1976. During this period, Captain Bausch's professionalism, leadership, and ingenuity were instrumental in Department of Defense preparation for the utilization of an advanced imagery collection system. With this experience added to his comprehensive knowledge of the collection field, he aggressively espoused a multisource concept for Department of Defense intelligence collection management and successfully engineered the transition of the new Department of Defense Collection Coordination Facility from a conceptual phase to an operational entity. His contribution to the Directorate for Collection Operations and the Defense Intelligence Agency contributed immeasurably to the command's mission. By his distinctive achievements, perceptive judgment, and steadfast dedication, Captain Bausch reflected great credit upon himself and upheld the highest traditions of the United States Naval Service.

For the President,

J. William Middendorf

Secretary of the Navy

Appendix B–Other Related Documents/Photos

Commission-Ensign USNR (May 1955)

THE

President of the United States of America.

To all who shall see these presents, greeting:

Know Ye, that reposing special Trust and Confidence in the Patriotism, Valor, Fidelity and Abilities of FRANCIS ADAM BAUSCH *I do appoint him* ENSIGN *in the Naval Reserve of The United States Navy to rank from the* SIXTH *day of* MAY, 1955. *He is therefore carefully and diligently to discharge the duties of such office by doing and performing all manner of things thereunto belonging.*

And I do strictly charge and require all Officers, Seamen and Marines under his Command to be obedient to his orders. And he is to observe and follow such orders and directions from time to time as he shall receive from me, or the future President of The United States of America, or his Superior Officer set over him, according to the Rules and Discipline of the Navy.

This Commission to continue in force during the pleasure of the President of the United States, for the time being.

Done at the City of Washington this FOURTEENTH *day of* FEBRUARY *in the year of our Lord One Thousand Nine Hundred and* FIFTY-FIVE *and of the Independence of The United States of America the One Hundred and* SEVENTY-NINTH.

By the President:

C. S. Thomas

Secretary of the Navy

669808

VC-61 Newsletter Excerpts

9 February 1956: A new officer reported in this week and was introduced in the ready room as Ensign Frank Bausch. There was a polite round of greetings until LT Ted Daum casually remarked, "Frank's our new P.I." Five OINCs of five upcoming detachments collided violently in their race to shake Frank's hand and inquire if he liked sea duty. For background information, I might say Frank's a bachelor from New York City who heeded Horace Greeley's advice on going west. As a P.I. among the vultures, he's about half-way West right now. Frank further torpedoes the hopes of the senior officer to wax the junior officer in basketball as he was a member of one of those fine St. John's team.

23 March 1956: Detachment Yankee leaves VC-61 next week for some strange ports of call. Lt. Ralph and intrepid PI ENS Frank Bausch will make up the team when they headquarter at Alameda for the better part of a week to clean up the myriad of special photo projects in the northern area. Needless to say, Ralph will fly and Frank will interpret and lay the mosaics. It should be interesting to see who does the most work. *(Ralph ran a tank empty in the Beechcraft over the San Gabriels going up–result was a scary deafening silence)*.

ENS Frank Bausch was drinking a cup of coffee in the ready room the other afternoon after a morning of slaving over hot mosaics in PI. Over the squawk box came this word: "Mr. Bausch, your aircraft is ready." "MY WHAT?" "Your F9F-5P, sir.". Frank has been bumming hops so regularly in recent weeks the line crew had him on the pilots' list! (P.S. I think if the plane hadn't gone down, he'd have taken it.)

18 May 1956: Ens. Frank Bausch, the Hotel PI, is getting closer to aviating every day. This week he moved up to two hops in the Cougar OFT. He's mastered those early "uncontrollable spins," and he does right well on normal maneuvers. If enthusiasm ever made a pilot, Frank would be one.

C-119 Flying Box-Film Capsule Catch

NAVPIC Open House-1960

U.S. ATLANTIC FLEET

ANTI-AIR WARFARE TRAINING CENTER

This is to certify that

LCDR. F. A. BAUSCH

has successfully completed the

Electronic Warfare Officers Course

This 18th day of FEB 1966

Training Officer

Commanding Officer

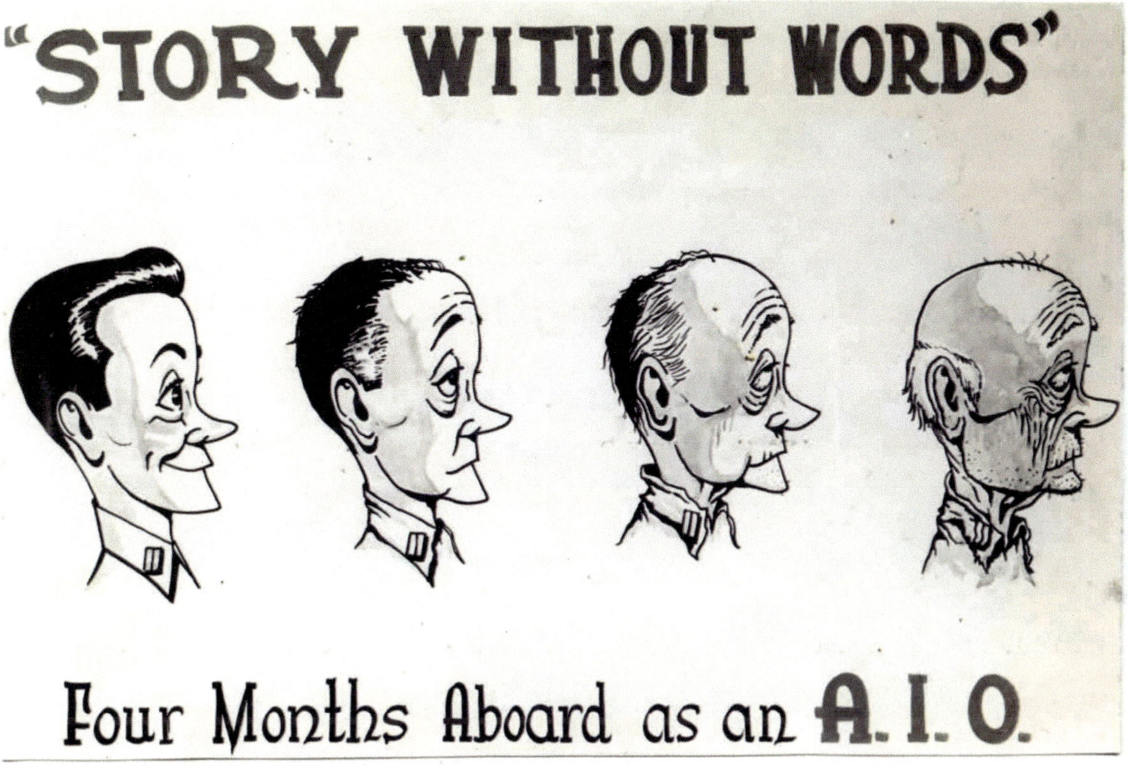

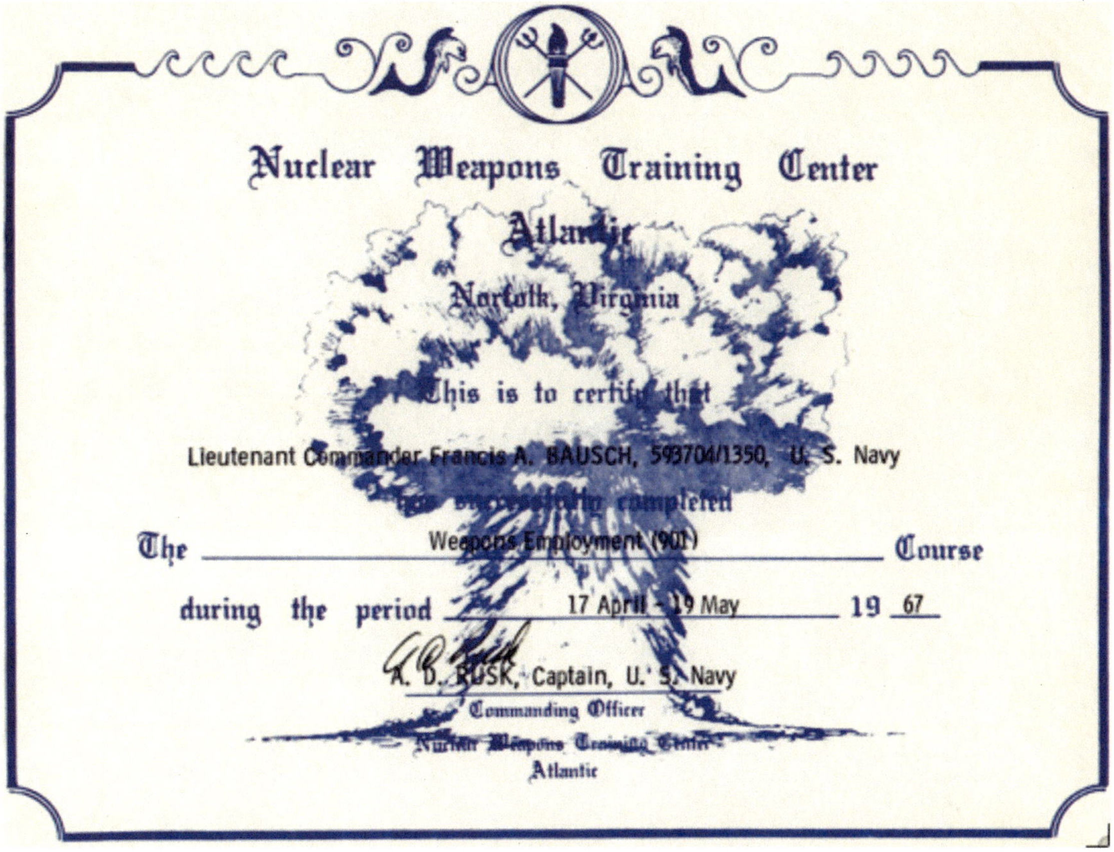

Nuclear Weapons Training Center
Atlantic
Norfolk, Virginia

This is to certify that

Lieutenant Commander Francis A. BAUSCH, 593704/1350, U. S. Navy

has successfully completed

The Weapons Employment (901) Course

during the period 17 April - 19 May 19 67

A. D. RUSK, Captain, U. S. Navy
Commanding Officer
Nuclear Weapons Training Center
Atlantic

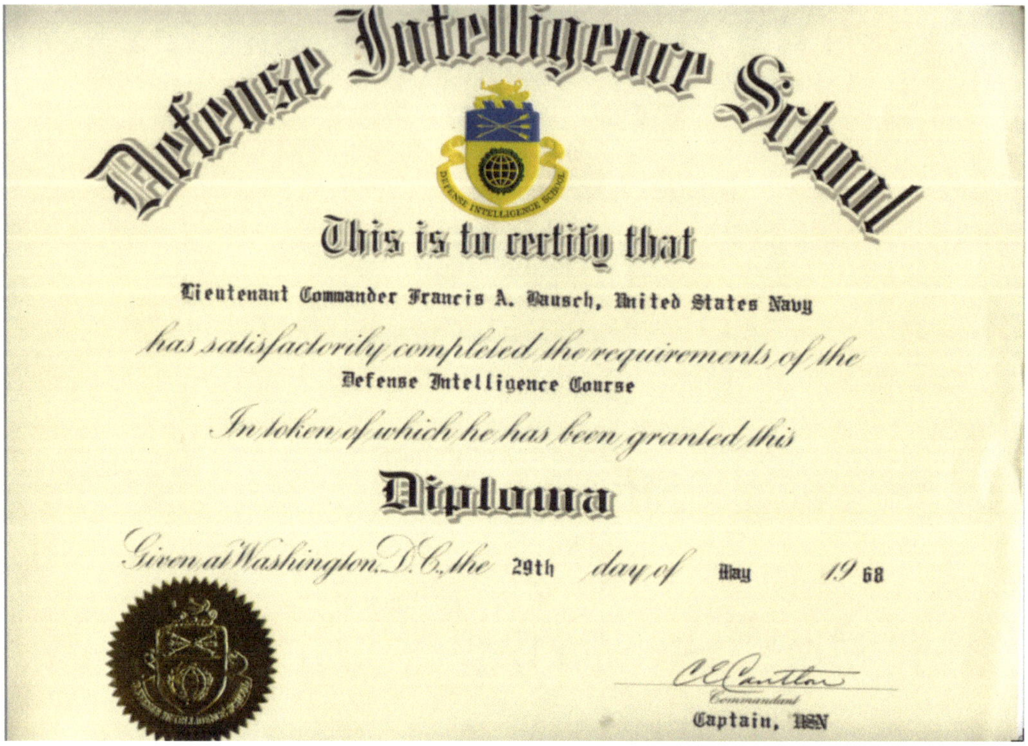

Defense Intelligence School

This is to certify that

Lieutenant Commander Francis A. Bausch, United States Navy

has satisfactorily completed the requirements of the

Defense Intelligence Course

In token of which he has been granted this

Diploma

Given at Washington, D.C., the 29th *day of* May *19*68

Commandant

Captain, USN

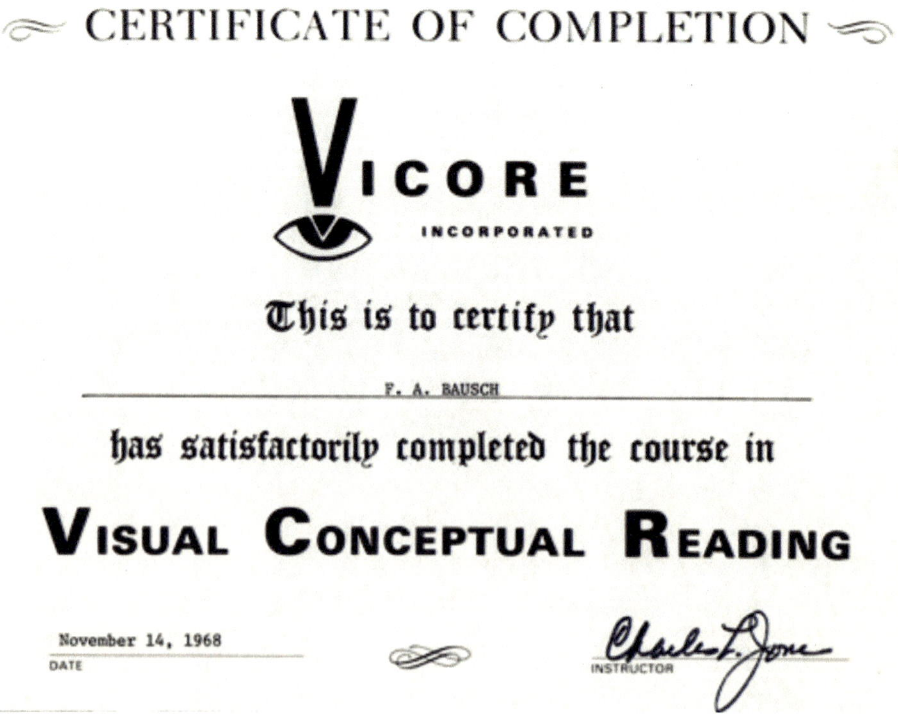

CERTIFICATE OF COMPLETION

VICORE

INCORPORATED

This is to certify that

F. A. BAUSCH

has satisfactorily completed the course in

VISUAL CONCEPTUAL READING

November 14, 1968

DATE

INSTRUCTOR

DNRO July 6, 1967 Memo to DepSecDef (extract)

4. OXCART Program: As a result of a BOB/DOD/CIA study last year, a decision was made by the President to phase-out the OXCART program during FY 1968. Initial phase-out of OXCART aircraft was to have been initiated in July 1967 with the mothballing of two aircraft culminating a complete phase-out by December 1967. The OXCART contingency plans, SKYLARK (Cuban operations in the event of hostile action against U-2s) and BLACK SHIELD (deployment of three aircraft to Kadena for operations in S. E. Asia and China), were to have been taken over by Air Force SR-71s. However, in May of this year the President directed the BLACK SHIELD deployment but without authorizing overflights of China. As a result the CIA OXCART program office submitted to me two options for modifications of the phase-out plan. The first would have delayed all phase-out until December 1967 and the second would have delayed phase-out throughout FY 1968 with a decision to continue beyond that left open. I recommended the first option to Mr. Vance. After consultation with Mr. Helms and Mr. Schultze of BOB, he directed the modification of the original phase-out plan only to the extent of retaining one of the two aircraft scheduled to be mothballed in July 1967. One problem with the SR-71 takeover of OXCART responsibilities is the delay in achievement of satisfactory ECM capabilities. We now have expedited the ECM program to achieve three operationally suitable SR-71 aircraft in September.

The approved cover story for the BLACK SHIELD deployment to Kadena is that the aircraft are "experimental test bed versions of the YF-12A/SR-71 deployed to Kadena for field tests." As was anticipated, after two weeks of operation from Kadena the aircraft were noticed by the Okinawan press. The prepared cover story was given to them with a "no comment" reply to all the additional questions which naturally ensued. Mr. Vance has expressed concern over the possible allegations of lack of credibility which may be attached to this operation and cover story. He has asked Mr. Katzenbach to query our Ambassador to Japan, Mr. Johnson, on the advisability of openly announcing that these aircraft are engaged in reconnaissance over North Vietnam. While this in itself would not be a surprising announcement in view of all our other reconnaissance activities over Vietnam, the CIA involvement in this program does pose special problems, and I am certain that their reaction to such an approach would be one of concern.

5. TAGBOARD Drones: This Mach 3, 90,000 ft altitude drone program is largely supported by the DOD in the ExCom. The CIA has always been lukewarm toward drone programs; Mr. Vance on the other hand has always been a strong adherent of doing reconnaissance of denied areas unmanned if at all possible. In the FY 1968 budget review, the BOB recommended that this program be slipped in schedule and reduced in funding or cancelled outright. Mr. Vance made this an issue for the President, who decided in favor of continuing the program. Although there were two reasonably successful flights of this drone under the original concept of launching from an OXCART, an accident resulting in the loss of a launch aircraft resulted in a re-evaluation of the program and a decision to convert to launch from a B-52 with rocket assist of the drone to altitude. The first launch from B-52 is to occur in August. The problem area at present is in qualifying the booster rocket.

CIA/Air Force NRO Agreement (extract)

ILLEGIB

ANNEX A.

The following assignments for the development of new optical sensor subsystems are made to take full advantage of technical capability and experience of the agencies involved.

1. The CIA will develop the improvements in the CORONA general search optical sensor subsystems.

2. Following the selection of a concept, and a contractor, for full-scale development, in the area of advanced general search, the CIA will develop the optical sensor subsystem for that system.

The Director, NRO will, in managing the corresponding over-all systems developments, ensure that:

1. The management of and contracting for the sensors is arranged so that the design and engineering capabilities in the various contractors are most efficiently utilized.

2. The sensor packages and other subsystems are integrated in an over-all system engineering design for each system, with DNRO having responsibility for systems integration of each over-all system.

25X1

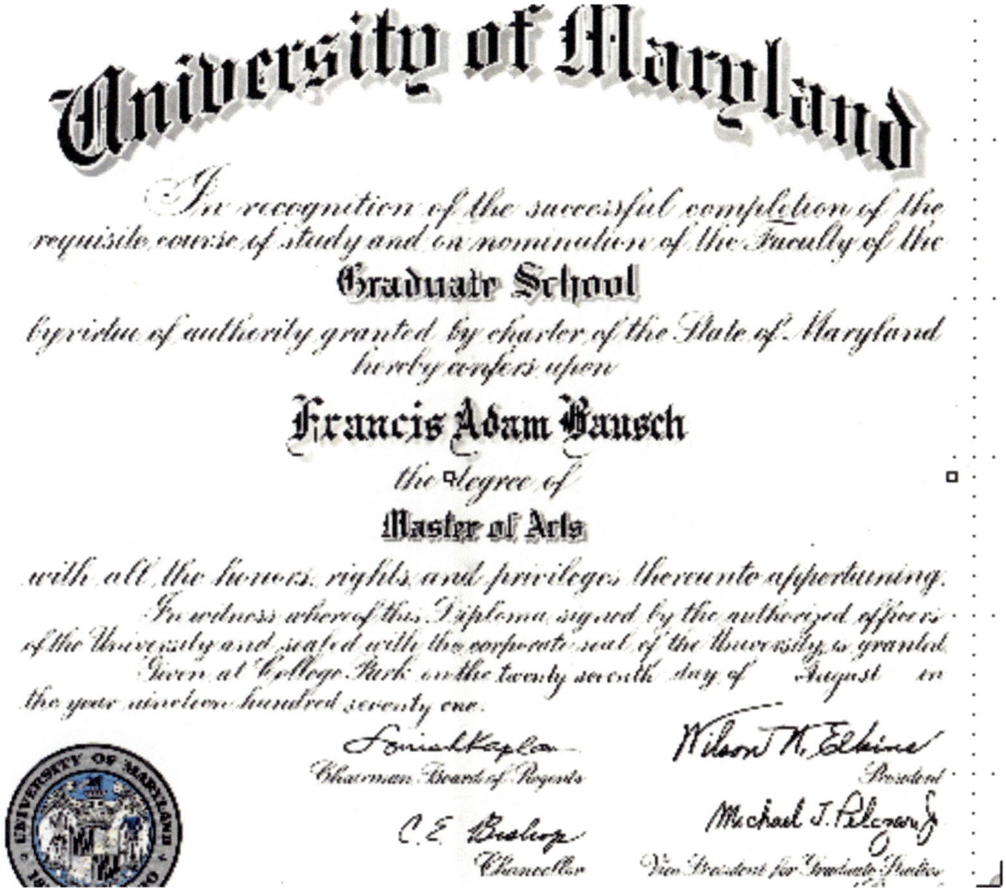

University of Maryland

In recognition of the successful completion of the requisite course of study and on nomination of the Faculty of the

Graduate School

by virtue of authority granted by charter of the State of Maryland hereby confers upon

Francis Adam Bausch

the degree of

Master of Arts

with all the honors, rights and privileges thereunto appertaining. In witness whereof this Diploma signed by the authorized officers of the University and sealed with the corporate seal of the University is granted. Given at College Park on the twenty seventh day of August in the year nineteen hundred seventy one.

Chairman Board of Regents

Wilson H. Elkins
President

C.E. Bishop
Chancellor

Michael J. Pelczar Jr.
Vice President for Graduate Studies

NIOC Briefing Slide (1971)

Naval Intelligence Officer POI Cover

POI 3OBR8051-4
(PDS Code QCK)

PLAN OF INSTRUCTION

(Technical Training)

NAVAL INTELLIGENCE OFFICER

AIR TRAINING COMMAND

LOWRY TECHNICAL TRAINING CENTER
15 January 1973 - Effective 20 January 1973

FOR OFFICIAL USE ONLY

Navy
Lowry AFB, Colorado
Staff Officers, Chief Petty Officers, and Non-Commissioned Officers
1 January 1972
CDR F. A. Bausch
Officer in Charge, Naval Unit and Deputy Director, AFAITC
LCDR B. Mater
LCDR W. Halloran
LCDR F. Skinner
LCDR C. Baker
MAJ R. Pelott
LCDR W. Crossman
LT E. Johnson
LT R. Thompson
LT D. Hayzlett
LT D. Foote
LT J. Kramer
LT K. McGovern
LT L. Joslin
LT M. Heffernan
LT D. Joslin
LT D. Peterson
LT C. Humphrey
LT J. Goss
LT L. Jensen
CWO3 S. Sanders
ETCM E. Fithian
YNCS J. Helmick
MSGT R. Moore
PTC J. Kimball
GYSGT M. Chesney
PTC G. Gass
PTC J. Rowe

COLORADO SEMINARY
UNIVERSITY OF DENVER
UNIVERSITY PARK • DENVER, COLORADO 80210

1600
1415
CLANCY

THE GRADUATE SCHOOL OF INTERNATIONAL STUDIES / THE SOCIAL SCIENCE FOUNDATION / 303 • 753-2324

March 29, 1972

Cmdr. Francis A. Bausch
11671 East Colorado Drive
Aurora, Colorado 80010

Dear Commander Bausch:

I am pleased to inform you that you have been accepted for admission to the Graduate School of International Studies for the 1972-73 academic year.

We hope that you will accept our offer of admission. We would enjoy having you as a member of the student body of our School.

We would appreciate it if you would notify us by telephone not later than April 7th of your decision to accept or reject this offer. Alternate candidates for admission awards have been named. So that they will not be kept in a state of uncertainty indefinitely, awards will pass to alternates if prime candidates have not responded by the above date.

If you have any questions about our program or about this award, please feel free to call Professor David H. Bayley, Chairman of the Admissions Committee. He can be reached at 753-2988.

Congratulations and best wishes!

Sincerely yours,

Robert Good

Robert C. Good
Dean

DWB:ac

JCS 2150/400-2

18 March 1974

COPY NO. ______

DISTRIBUTION C

OPI CF

SC
ACTION
VAULT

CS

DISSEMINATION OPTIONS

DECISION

1. At their meeting on 20 March 1974, the Joint Chiefs of Staff, after making an amendment, approved the recommendations contained in paragraph 8 of this report.*

2. This decision replaces page 3. Holders are requested to destroy the superseded page in accordance with security regulations.

3. The memorandum in Enclosure A, together with its Appendices and Annexes, was forwarded as JCSM-94-74,** dated 22 March 1974, to the Secretary of Defense. Copies of this paper* are being forwarded to CINCLANT, USCINCEUR, CINCPAC, and CINCSAC.

4. The Commandant of the Marine Corps expressed the direct concern of the Marine Corps in this matter; accordingly, the provisions of Title 10, US Code 141 (c), applied and were followed.

DISTRIBUTION:

Adm Moorer (C/JCS)	(2)	Adm Boyd (DD/JS)	(1)
Gen Abrams (CSA)	(1)	Adm King (J-3)	(2)
Adm Zumwalt (CNO)	(2)	Gen Seith (J-5)	(2)
Gen Brown (CSAF)	(1)	Gen Schulke (J-6)	(2)
Gen Cushman (CMC)	(1)	Gen Gould (DCA)	(1)
Gen Cowles (DC/S, OPS)	(5)	Adm de Poix (DIA)	(9)
Adm Talley (DCNO-P&P)	(4)	Gen Allen (NSA/CSS)	(1)
Gen Huyser (DC/S, P&O)	(5)	Gen Penney (DMA)	(1)
Gen Beckington (DC/S, P&O, MC)	(3)	Gen Wilcox (S/JCS)	(1)
Gen Seignious (D/JS)	(1)	Col Jones (DS/JCS)	(4)

Action Officer: Cdr F. A. Bausch, USN,
DIA, Ext 57659

* Available in SAO channels; cite no. TCS-P-565479/74
** Available in SAO channels; cite no. TCS-P-565561/74

JCS 2150/400-2 3

(Revised - 22 March 1974)

RETAIN OR DESTROY

PRESIDENT
OF
THE UNITED STATES OF AMERICA

To all who shall see these presents, greeting:

Know ye that, reposing special trust and confidence in the patriotism, valor, fidelity, and abilities of FRANCIS A. BAUSCH *, I do by and with the advice and consent of the Senate, appoint* HIM A

CAPTAIN

United States Navy

to rank as such from the FIRST *day of* SEPTEMBER *, nineteen hundred and* SEVENTY-FIVE *. This officer will therefore carefully and diligently discharge the duties of the office to which appointed by doing and performing all manner of things thereunto belonging.*

And I do strictly charge and require those officers and other personnel of lesser rank to render such obedience as is due an officer of this grade and position. And this officer is to observe and follow such orders and directions, from time to time, as may be given by the President of the United States of America, or other superior officers, acting in accordance with the laws of the United States of America.

This commission is to continue in force during the pleasure of the President of the United States of America, under the provisions of those public laws relating to officers of the **Armed Forces of the United States of America** *and the component thereof in which this appointment is made.*

Done at the City of Washington, this TWENTY-SECOND *day of* SEPTEMBER, *in the year of our Lord, one thousand nine hundred and* SEVENTY-EIGHT *, and of the Independence of the United States of America, the* TWO HUNDREDTH AND THIRD .

By the President:

DEPARTMENT OF THE NAVY

The National Cryptologic School

National Security Agency

Awards this certificate to

CDR FRANCIS A. BAUSCH, USN

in recognition of satisfactory completion of

CRYPTOLOGIC FAMILIARIZATION COURSE FOR SENIOR DEFENSE OFFICIALS

conducted by the National Cryptologic School

DATE: 21 November 1974

LEW ALLEN
Lieutenant General, USAF
Director
National Security Agency

THOMAS A. PRUGH
Commandant
National Cryptologic School

Chick DeThomasis, SecNav Middendorf, FAB, SecAF McLucas

DEPARTMENT OF THE NAVY
BUREAU OF NAVAL PERSONNEL
WASHINGTON, D.C. 20370

IN REPLY REFER TO

N0002277T07ZU08
067-26-0478/1630
PERS-472-wb-2
23 NOV 1976

BUPERS ORDER
T-01785

CAPTAIN FRANCIS A. BAUSCH, USN
BUPERS

COMMENCING THIS DATE WHEN VERBALLY DIRECTED FROM TIME TO TIME, PROCEED SUCH PLACES INSIDE AND OUTSIDE CONUS INCLUDING PORTUGAL AS NECESSARY TEMADD CONNECTION NAVAL MATTERS.

{1} {2} {3} {4} {5} {6} {7} {8} {9}
COST TVL CHG AA 1771804.2280 005 00022 Z 065872 2D 70ZU08 009602U08XZE. CIC 377ZU08N00022AA. EST COST TRANS $1,500.00, PER DIEM $300.00, MISC $200.00.

ITEM 29; 31; 201(A); 204; 210; 211; 213; 216(2); 222; 225; 230-NAVMATCOMSUPPACT, WASHINGTON, DC (SA 123)

UNDER CIRCUMSTANCES NOT PERMITTING TRAVEL BY THE USUAL MEANS OF TRANS BETWEEN AND IN AND AROUND TEMADD POINTS, YOU ARE AUTHORIZED TO HIRE A GSA CONTRACT AUTOMOBILE IF AVAILABLE, OTHERWISE COMMERCIAL AUTOMOBILE FOR OFFICIAL BUSINESS IN ACCORDANCE WITH JTR, PAR M4405-2-3-5 AND PAR M4501, ITEM 3.

YOU ARE AUTHORIZED THE USE OF TAXICABS IN AND AROUND TEMADD POINTS FOR OFFICIAL BUSINESS IN ACCORDANCE JTR, PAR M4413-3 AND PAR M4501 ITEM 2.

AUTHORITY TERMINATES 30 SEP 77.

YOU SHALL PAY CASH FOR CAR RENTAL SERVICES.

LA PERSONA MENCIONADA EN ESTOS DOCUMENTOS HA SIDO AUTORIZADA POR EL COMANDANTE DE LA UNIDAD A LA QUE PERTENECE PARA DESPLAZARSE A ESPANA.

FINAL TOP SECRET CLNC BASIS BI COMPLETED 14 SEP 76 BY DIS.

James D. Watkins

JAMES D. WATKINS
VICE ADMIRAL, USN
CHIEF OF NAVAL PERSONNEL

JACKET
PERS-475 131 472

CERTIFICATE OF RETIREMENT

FROM THE ARMED FORCES OF THE UNITED STATES OF AMERICA

TO ALL WHO SHALL SEE THESE PRESENTS, GREETING:

THIS IS TO CERTIFY THAT

CAPTAIN FRANCIS A. BAUSCH

HAVING SERVED FAITHFULLY AND HONORABLY

WAS RETIRED FROM THE

UNITED STATES NAVY

ON THE FIRST DAY OF JANUARY

ONE THOUSAND NINE HUNDRED AND SEVENTY-NINE

DEPARTMENT OF THE NAVY · UNITED STATES OF AMERICA

Robert B. Baldwin

VICE ADMIRAL, UNITED STATES NAVY
CHIEF OF NAVAL PERSONNEL

Made in the USA
Monee, IL
15 November 2026

68086625R10079